Workshop on Contemporary Materials and Technologies in the Aviation Industry (CMTAI 2016)

MATEC Web of Conferences Volume 99 (2017)

Moscow, Russia
15 - 16 December 2016

Editors:

E.V. Nikulchev
O.N. Korsun

ISBN: 978-1-5108-3609-9

Printed from e-media with permission by:

Curran Associates, Inc.
57 Morehouse Lane
Red Hook, NY 12571

Some format issues inherent in the e-media version may also appear in this print version.

Printed by Curran Associates, Inc. (2017)

For additional information, please contact EDP Sciences – Web of Conferences at the address below.

EDP Sciences – Web of Conferences
17, Avenue du Hoggar
Parc d'Activité de Courtabœuf
BP 112
F-91944 Les Ulis Cedex A
France

Phone: +33 (0) 1 69 18 75 75
Fax: +33 (0) 1 69 28 84 91

contact-edps@webofconferences.org

Additional copies of this publication are available from:

Curran Associates, Inc.
57 Morehouse Lane
Red Hook, NY 12571 USA
Phone: 845-758-0400
Fax: 845-758-2633
Email: curran@proceedings.com
Web: www.proceedings.com

TABLE OF CONTENTS

Aims & scope

MATEC Web of Conferences is an open-access journal devoted to the publishing of conference proceedings dealing with Materials science, Engineering and Chemistry.

The journal aims to make conference proceedings visible, citable and widely usable by the scientific community. It offers outstanding services for the publication and dissemination of contents in order to ensure long-term archiving and maximum exposure of scientific events.

Coverage includes all engineering disciplines: civil, naval, mechanical, chemical, and electrical engineering as well as nanotechnology and metrology. The journal also deals with the physical-chemical characterization of materials, their implementation, their resistance in their environment, etc. Others subdisciplines of chemistry, such as analytical chemistry, petrochemistry, organic chemistry, etc., and even pharmacology, are also welcome.

Why publish with MATEC Web of Conferences?

- **Open access:** conference papers available worldwide.
- **Fast publication:** less than 3 months.
- **Total flexibility:** no page restrictions, free unlimited use of colour in the online version, any kind of scientific document welcomed (traditional articles, posters, abstracts and slideshows).
- **High visibility:** all documents indexed in main databases and easily found online.
- **Fair author rights:** CC-BY Creative Commons license.
- **Range of additional publishing options:** printed proceedings, CD-ROMs, copy-editing, etc.

ISSN (electronic edition): 2261-236X

Contact: Isaline Augusto
matec@webofconferences.org

2016 Workshop on Contemporary materials and technologies in the aviation industry (CMTAI2016)

Moscow (Russia), 15–16 December 2016

Main topics:

- Composite materials & structure development
- Aircraft engines
- Aerodynamics Technology
- Civil Aviation Technology
- Guidance, Navigation and Control

Scientific Committee:

Members represent institutes and universities:

- Bauman Moscow State Technical University, 105005 Moscow, Russia
- Branch of JSC „SRC „Progress" – Special Design Bureau „Spectr", Ryazan, Russia
- FGUP "GosNIIAS", Moscow, Russia
- Institute of Chemistry and Chemical Technology Russian Academy of Sciences, Krasnoyarsk, Russia
- JSC "Spatial information systems", Moscow, Russia
- Kuban State University, 350040 Krasnodar, Russia
- Moscow Aviation Institute, Aeronautical school, Moscow, Russia
- Moscow Institute of Physics and Technology, Dolgoprudny, Russia
- Moscow Polytechnic University, Moscow, Russia
- Moscow Scientific-Research Television Institute, Moscow, Russia
- Moscow State Technical University of Civil Aviation, Moscow, Russia
- Moscow Technological Institute MIREA, Moscow, Russia
- Moscow Technological Institute, Moscow, Russia
- National Research University Higher School of Economics, Moscow, Russia
- People's Friendship University of Russia, Moscow, Russia
- Ryazan State Radio Engineering University, Ryazan, Russia
- Samara state technical university, Samara, Russia
- Southwest State University, Kursk, Russia
- University MSBY of Jijel, Jijel, Alegria
- Uspensky Avionica Moscow Research and Production Complex, Moscow, Russia
- V.A. Trapeznikov institute of Control Science of Russian Academy of Sciences, Moscow, Russia

edp sciences

MATEC Web of Conferences

Statement of Peer review

In submitting conference proceedings to MATEC Web of Conferences, I certify to the Publisher that I adhere to the **Policy on Publishing Integrity** of the journal in order to safeguard good scientific practice in publishing.

1. All articles have been subjected to peer review administered by the proceedings editors.
2. Reviews have been conducted by expert referees, who have been requested to provide unbiased and constructive comments aimed, whenever possible, at improving the work.
3. Proceedings editors have taken all reasonable steps to ensure the quality of the materials they publish and their decision to accept or reject a paper for publication has been based only on the merits of the work and the relevance to the journal.

Title, date and place of the conference

2016 Workshop on Contemporary materials and technologies in the aviation industry (CMTAI2016)

Moscow, Russia, 15–16 December 2016

Proceedings editor(s):

E.V. Nikulchev, professor
O.N. Korsun, professor

Date and editor's signature

01/27/2017

DOI: 10.1051/matecconf/20179901001

Materials for Test Objects to Configure Aviation Optoelectronic Earth Remote Sensing Systems

E.B. Bablyuk[1], *Yu. M.* Berlad[1], *A.G.* Letyago[2] and *A.P.* Kondratov[1,*]

[1]Moscow Polytechnic University, 107023 Moscow, Russia
[2]Moscow Scientific-Research Television Institute, 105094 Moscow, Russia

Abstract. The paper illustrates the possibility of manufacturing deployable test objects to configure aviation optical equipment for remote sensing of the Earth in the form of a set of fragments made of weatherproof polymeric materials using the method of digital inkjet printing of two types of prints. The main fragments with a fixed optical density include images of light bars against a dark background. Background fragments are monochrome. We have suggested a reasonable combination of materials for test objects fragments, and assessed their optical parameters and lightfastness.

1 Introduction

Analogous optoelectronic systems for monitoring of the Earth's surface (MES) are being replaced by MES digital optoelectronic systems (OES). Unlike film systems, digital OES uses matrices of elementary photosensitive elements with a strictly ordered regular structure as an optical emission receiver (OER). Used as an optical emission receiver of digital OES, matrix (linear) charge-coupled photosensitive devices (CCPD) or metal-oxide-semiconductor complementary structures (MOSCS) have a fixed the physical size of an OER element. For this reason, if we want to assess the resolution power and to configure digital optoelectronic MES systems, it is necessary to create and to use bar test objects with a certain angular orientation of strips (bars) with respect to the regular structure of the optical emission receiver.

As the key MES OES performance evaluation criteria, the linear ground resolution of the onboard system $L_м$ is used.

The linear ground resolution is determined using the following formula:

$$L_м = \frac{H}{2Rf},$$

where R – is the system's resolution power, mm^{-1}; H is the height of the aerial survey, m; f is the system's focal distance, mm.

The resolution power is a spatial-frequency characteristic of a photographic system. It is determined by visual decoding after photographing the test objects. The resolution power of MES OES is determined both in laboratory and in flight environments during navigation tests.

Configuration kits for airfield-based aviation optoelectronic systems include test objects containing calibrated photometric fields [1]. The spatial resolution of aviation equipment is assessed using bar test objects. Radiometric properties are assessed using photometric fields. There are two types of test objects: fixed objects manufactured by applying paint on a concrete base (the paint composition includes additives to ensure the required surface diffuseness) and deployable objects – banners made of fabric or other flexible material with a printed pattern.

Test objects are rather material-intensive and shall have stable measuring properties throughout their service life. Their reasonable designing, manufacturing, deployment and maintenance of the surface in working condition shall meet high standards [1+]. Not only the choice of material for a test object is important, but also the technological process of printing the bars and bright fields, to ensure a high accuracy and stability of the pattern over large areas of the printed material.

The purpose of this study is justifying the optimum design, the choice of materials and the assessment of the bars and bright fields printing quality on a deployable test objects used to configure aviation optoelectronic system for remote sensing of the Earth.

2 Research Objects and Methods

To investigate the prints' optical properties, we selected banner materials of the "BlockOut" trademark, with a PVC coating.

Surface reflection spectral coefficients were measured using a "FieldSpec 3" spectroradiometer. The reflection coefficients were determined on a blank banner material before printing and on bright fields and bar elements with a variable tone intensity (30%, 55% and 100%).

Before each measurement, we determined the optical parameters of each white master object. The position

* Corresponding author: apkrezerv@mail.ru

height of the spectroradiometer's fiber optic cable lens was 15 cm.

To assess the lightfastness, we used a print on the "Blockout" S-PRW334240B banner material.

The lightfastness was assessed by measuring the optical density of printed areas with a tone intensity of 55% and 100% after exposure of samples in a device according to the patent [6].

3 Results and Discussion

The properties analysis of weatherproof printed materials, that are available on the world market [5–7], allows us to conclude that banner fabrics with a PVC coating are suitable to manufacture test objects for remote sensing of the Earth. Synthetic fabric banners are

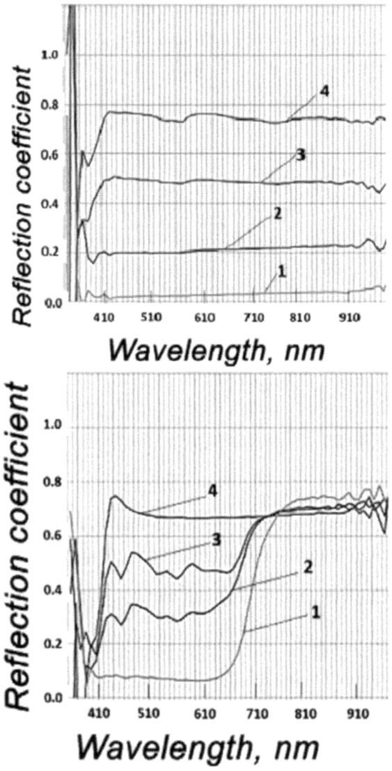

Fig. 1. The spectrum of light reflection coefficients from the prints on the fabric sample № 1 (a) and the fabric sample № 2 (b) with the tone intensity of 100% – 1, 55% – 2, 33% – 3, 1% – 4.

laminated composite materials that consist of a polyester mesh and a monolithic polyvinyl coating, intended for printing on polygraphic equipment. In terms of banners manufacturing technology, the mesh is filled with a solution or pasty composition (plastisol), containing polyvinyl chloride, and then it is passed through hot press rolls for imparting surface smoothness to the material, as required for printing.

The printed material for test objects shall have lightfastness, the minimum light transmittance and uniform electromagnetic radiation reflection characteristics in the visible spectrum before and after printing patterns (bars) with variable optical density thereon.

Based on a comparative evaluation of the physical and mechanical properties values, as declared by the manufacturer [7], and specific quality characteristics of the printed surface for manufacturing test objects and subsequent tests, we selected the following fabric banners: "BackLit" GLK S5251, "BlockOut" GLK S6151, "BlockOut" S-PRW334240B, "BlockOut" GLK S5355M, as well as and nonwoven fabrics made of synthetic fibers.

The values of the spectral reflection coefficients are presented in the form of charts (Fig. 1).

The light reflection coefficient from the non-printed fabric banner № 1 (1%) within the range of 450–1,000 nm has a constant value. The spectrum of light reflection coefficients from the prints with a tone intensity of 30% and 55% within the range of 400–600 nm is unstable. All prints with a tone intensity of (30%, 55%, and 100%), that were obtained on the fabric banner sample № 2 (Fig. 1, b), have a constant value of light reflection coefficient in the wavelength range of 450–1,000 nm.

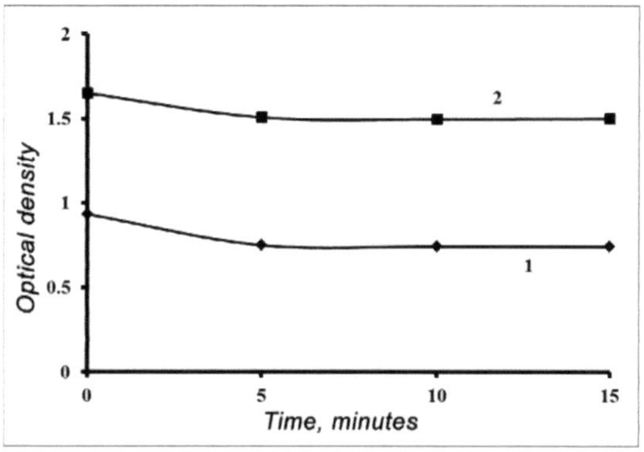

Fig. 2. Changes in the optical density of ink layer on different materials after printing using an UV ink: 1 – nonwoven material; 2 – banner fabric.

Modern banner inks are UV-curable, i.e. light-absorbing compositions that significantly change their state of aggregation and physical properties when exposed to short-wave electromagnetic radiation. Therefore, it is reasonable to assume that during manufacture, immediately after printing and during a long-term use of test objects, the stability of their optical properties is likely to be affected not only by ultraviolet radiation, but also by daytime sunlight.

It has been found that the optical density of prints depends on the type of the non-printed fabric and varies significantly during 5 minutes (Fig. 2), whereas the color difference of the prints, as obtained on the banner fabric, exceeds the sensitivity threshold of human eyes, i.e. it is can be observed visually.

For experimental evaluation of lightfastness, we studied a print made on the "Blockout" S-PRW334240B banner fabric.

The lightfastness was evaluated by measuring the optical density of printed areas with a tone intensity of 55% and 100% after aging the samples in a climatic chamber that simulated an accelerated deterioration of the material under the action of sunlight [6].

The testing results are presented in Table 1. The test object according to the technical solution [7] consists of fragments with an image of bars of a certain group and unified lateral background fragments of the same geometric dimensions. The fragments are made of a banner fabric using a printing method that ensures the required optical measurement accuracy.

Table 1. Optical density values of the prints on the "Blockout" S-PRW334240B banner material before and after exposure to UV radiation

Exposure time, hour	0		96		180	
Tone intensity, %	55	100	55	100	55	100
Optical density	0.848	1.226	0.852	1.228	0.846	1.224

No less important task, besides increasing the accuracy of the measurements, shall be reducing the costs of consumables for manufacture of test object, saving time and labor costs as required for modules deployment to test specific operation modes of onboard optoelectronic equipment. The design of test objects provides for laying a set of modules, optimal for each type of the configured optoelectronic systems on the Earth's surface, consisting of one or more main fragments with a width that is the multiple of three meters, with groups of light bars against a dark background.

In determining the optimum set, it is necessary to take into account that the total width of the fragments formed in a module should exceed the width of the bar group deployed with respect to the long side of the module at an angle of 45° or 135°. The number of auxiliary unified lateral background fragments, that are adjacent to the main fragments, should ensure at least two-fold excess of the distance from the bar pattern to the module edge.

The auxiliary background fragments, having similar characteristics (the same spectral reflection coefficient) against the background on the main fragments with bars, may be made of a nonwoven material that is less expensive, as compared to the banner fabric of the main fragments.

The evaluation of manufacturing costs for the existing [8] and assumed test objects is based on the comparison of the consumption of materials and related costs for manufacturing background fragments for one test object module with the same type of bars.

When manufacturing a test object, the manufacture of background fragments would need a printed material with a three-layer structure (including a reinforcing fabric and a lightfastness polymer coating) with a total surface area of 2.354 sq. m. When manufacturing background fragments of one test object module as per [7], a one-layer nonwoven material without coating is used, and the consumption of printed material for background fragments is 1.325 sq. m. The saving of the printed material is up to 1.029 sq. m, which is equivalent of 44% in terms of surface area, and significantly more in terms of cost.

The material's printing properties at background fragments differ substantially from those of banner fabrics with a PVC coating. The roughness of fabrics with a PVC coating is characterized by a value of 3–7 μm. Due to the high surface strength and the optimum surface roughness of banner fabrics, the adhesion of printing ink layers reaches 70.0±0.9 MPa. The adhesive strength of the printing ink layers, that is used for printing the background of nonwoven fabrics, is only 17.0±0.8 MPa. However, this value is sufficient for multiple bending and folding the background fragment prints into a roll during their storage and unfolding on the Earth's surface.

4 Conclusion

We have demonstrated a possible manufacturing method by digital inkjet printing and presented a reasonable combination of materials for deployment of test objects to configure aviation optical equipment for remote sensing of the Earth.

References

1. F. Goryl, A. Burini, Calibration Test Sites and Cal/Val Portal Renovation. IVOS 21 (University of Lethbridge, Canada, 11/13 August 2009)
2. H. Kipphan. Handbook of Print Media (Springer, Moscow State Printing University, 2003)
3. S.A. Shadrina, A.V. Churkin, *Polygraphy*, **9**, 17–20 (2014)
4. A.V. Churkin, Digital Printing. Key Methods and Equipment, In *Proceedings of the All-Russian conference – Digital Printing. Issues of Technique and Technology*, pp. 14–16 (2014)
5. A. Churkin, *Polygraphy*, **4**, 39-40 (2015)
6. A.P. Kondratov, A Device to Determine the Chemical Stability and Lightfastness of the Ink Printing on a Polymeric Package, Patent of the Russian Federation № 153923, IPC G01N 17/02, (2016)
7. E.V. Altukhov, E.B. Bablyuk et al. A Test Pattern for Parameters Configuration and Determination of Optoelectronic Systems, Patent of the Russian Federation № 153923, IPC G01M 11/02 (2016)
8. Yu. G. Veselov, V. V. Tikhonichev, A. Danilin, *Science and education*, **4** (2012)

DOI: 10.1051/matecconf/20179901002

Basic Model of Investigation Regularities of Fatigue Aging of Engineering Devices' Elements in Aerospace Industry

Alexander Chursin[1], *Alexander* Semenov[1, 2, *] and *Anna* Ostrovskaya[1]

[1]People's Friendship University of Russia, ul. Miklukho-Maklaya 6, Moscow, 117198 Russia
[2]Moscow Technological Institute, Leninsky prospect, 38a, 119334 Russia

Abstract. The article deals with the general approach to the problem of fatigue of materials and elements in aerospace industry. The mathematical model describing the effect is presented and recommended for practical use. For solving different problems with the fatigue of the materials we recommend the described model with the given approximation type

1 Introduction

Mechanical properties of materials in aerospace industry are determined mainly on the basis of standard testing of samples according to the Russian standard GOST 2860-65. The main experimental information of these tests includes various types of fatigue curves used to determine the extreme values considered as properties. According to the approach used in this article, these curves are classified as standard fatigue curves of metal. We will outline the basic methodology based on the principles in [1]. The importance of new methods and innovations in aerospace industry is described in [2].

2 Model of investigation regularities of fatigue aging

Figure 1 illustrates the fatigue curve from Russian standard GOST 2860-65, which recommends to consider the horizontal plateau as an upper limit of complete fatigue strength (or σ_{-1}). In Fig. 1:

$$\sigma_{max} = \sigma_m + \sigma_a, \qquad (1)$$

where σ_{max} is the maximum stress of fatigue cycle; σ_m is the static constituent of fatigue cycle, average cycle stress; σ_a is the amplitude of cycle stress. This Standard does not present any special recommendations on approximation of the experimental data. In aerospace industry it is convenient to apply approximation by using polynomial equations for certain particular cases or statistic processing of the experimental data.

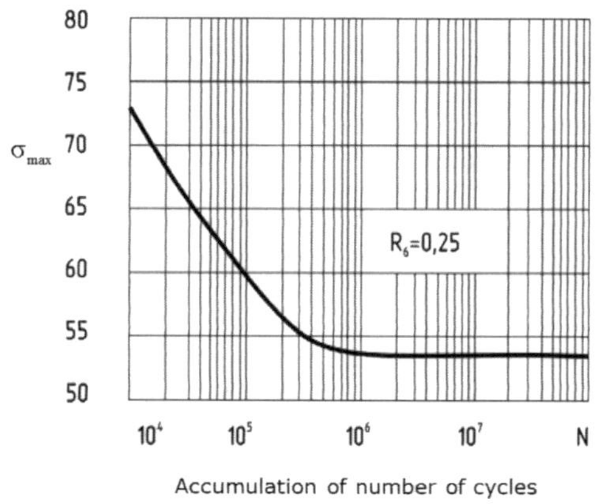

Fig. 1. Fatigue curve in coordinates σ_{max}, $\ln N$

Approximation of experimental results using transcendental equations makes it possible to relate main constants of material with probabilistic parameters of their expected behaviour upon aging.

Fig. 2 illustrates fatigue curve from Fig. 1 in Cartesian coordinates. The values of σ_{max} for $N_c < 104$ are extrapolated to $N_c \to 0$.

It is obvious that σ_{max} ($N_c = 0$) is nothing else but ultimate strength upon static load. In this case a sample fails in the first load cycle. In this plot it is the second constant, asymptote, approached by the curve $\sigma_{max} = f(N_c)$ at $N \to \infty$. Here and below the symbols σ_B, σ_R and others are applied in common sense without relation to the specific type of the load or deformation (twisting, bend, compression, expanding and so on).

For practical cases we recommend the approximation to be performed by the method of selected points. Four points indicated by numbers are

* Corresponding author: semenov.venture@mail.ru

 DOI: 10.1051/matecconf/20179901002

arbitrarily selected in the fatigue curve. Coordinates of these points are summarized in Tabl. 1 ($\sigma_R = 52.5$, $\sigma_B = 80$).

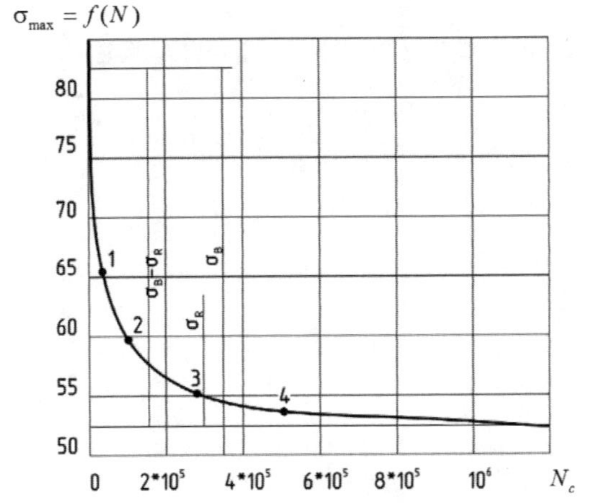

Fig. 2. Approximation of standard fatigue curve: σ_B — ultimate strength; σ_R — fatigue strength.

Table 1. Initial data

№ $_{TT}$	$N_{c\,i}$	σ_{max} (N_i)	№$_{TT}$	$N_{c\,i}$	σ_{max} (N_i)
1	0.3 10^5	65	3	2.9 10^5	55
2	1 10^5	59	4	5 10^5	53

Standard fatigue curve is described as follows:

$$\sigma_{max}(N) - \sigma_R = (\sigma_B - \sigma_R)e^{-\mu N \alpha}, \quad (2)$$

A special case of (2) at α=1:

$$\sigma_{max}(N) - \sigma_R = (\sigma_B - \sigma_R)e^{-\lambda N}. \quad (3)$$

In (2), (3) μ and α are the deterministic analogues of the Weibull–Gnedenko distribution parameters; λ is the intensity of decrease in strength (analogue of failure intensity). Transforming and applying the logarithm twice on (2) we obtain:

$$\lg\left[-\ln\left(\frac{\sigma_i - \sigma_R}{\sigma_B - \sigma_R}\right)\right] = \lg\mu + \alpha\lg N \quad (4)$$

Writing (4), for instance, for points 1 and 2 from Tabl. 1 and eliminating lg μ, we obtain:

$$\alpha = \frac{1}{\lg\frac{N_1}{N_2}} \cdot \lg\left[\frac{\frac{\sigma_1 - \sigma_R}{\sigma_B - \sigma_R}}{\frac{\sigma_2 - \sigma_R}{\sigma_B - \sigma_R}}\right].$$

For points 1 and 2:

$$\alpha = \frac{1}{\lg\frac{0{,}3 \cdot 10^5}{1 \cdot 10^5}} \lg\left[\frac{\frac{65-52{,}5}{80-52{,}5}}{\frac{59-52{,}5}{80-52{,}5}}\right].$$

For points 3 and 4:

$$\alpha = \frac{1}{\lg\frac{10^5}{2.9 \cdot 10^5}} \lg\left[\frac{\ln\left(\frac{59-52.5}{80-52.5}\right)}{\ln\left(\frac{55-52.5}{80-52.5}\right)}\right] = 0.4519$$

The results of calculations are loosely approximated, thus, α=0.45; and μ is determined by the following equation:

$$\mu = -\frac{\ln\left(\frac{\sigma_i - \sigma_R}{\sigma_B - \sigma_R}\right)}{N_i^{\alpha}} \quad (5)$$

For point 1 from Table 1:

$$\mu_1 = -\frac{\ln\left(\frac{65-52.5}{80-52.5}\right)}{(0.3 \cdot 10^5)^{0.45}} = 8.62 \cdot 10^{-3}$$

For point 2:

$$\mu_2 = -\frac{\ln\left(\frac{59-52.5}{80-52.5}\right)}{(10^5)^{0.45}} = 8{,}69 \cdot 10^{-3}$$

Calculation of the referential values by (5) gave nearly complete coincidence of the results.

3 Conclusions

Experimental application of (5) can be carried out by the following trends:

- selection of the load modes for a priori selected cumulative number for the load cycles NCyM. This procedure regulates selection of constrained load limit. Taking into account (1) and (2), the following equation can be derived:

$$\sigma_{amax}(N) + \sigma_m = \sigma_R + (\sigma_B - \sigma_R)e^{-\mu N\alpha},$$

where $\sigma_{amax}(N)$ is the ultimate value of load amplitude; σ_m is the static constituent, average cycle stress. Average cycle stress σ_m is selected a priori, and ultimate value $\sigma_{amax}(N)$ is calculated;

- forecasting of expected probabilistic properties, taking into account that the parameters μ, α, λ are identical to probabilistic dependences of the similar type.

So for solving different problems with the fatigue of the materials we recommend the described model with the given approximation type.

References

1. V. V. Efremov, V. A. Naumov, A. A. Chursin, *Theory and practical issues of operability of elements of machines, devices and instruments*, (Irkutsk University, Irkutsk, 1984) [In Rus]
2. A. A. Chursin, A. S. Semenov, A. V. Danilchankau, NPCS **19**, 3 (2016)

DOI: 10.1051/matecconf/20179901003

The Liquid Crystal State Poliamidbenzimidazola Solutions in Sulfuric Acid

Oleg Khanchich[1,*] *and Svetlana A.* Kuznetsova[2]

[1]Moscow Technological Institute, 119334 Moscow, Russia
[2]Institute of Chemistry and Chemical Technology Russian Academy of Sciences, 660036 Krasnoyarsk, Russia

Abstract. We studied the temperature and concentration conditions of education and the field of LC – phase of existence in sulfuric acid solutions poliamidbenzimidazola. The polarization–optical methods and the structural features of biphasic and anisotropic areas and built plots the phase diagram of the concentrated solutions poliamidbenzimidazola in H2SO4. It is shown that in certain temperature – concentration of cooling modes can be observed the coexistence of three phases: isotropic crystal and a liquid crystal, which is shown as a characteristic of liquid crystal birefringent domains.

1 Introduction

Currently, a wide importance in mechanical engineering, especially aircraft, purchased polymer composites. They are obtained on the basis of reinforcing fibers and polymer matrices [1]. Used as a template, along with other thermoplastics polyheteroarylenes. The best studied of these is poly-n-phenylene terephthalamide (PPTA) [2]. Fibers of this polymer is formed from a liquid crystal (LC) solutions in concentrated sulfuric acid. Less studied, but no less practical significance is poliamidbenzimidazol (PABI), which is also derived from the LC in H_2SO_4 solutions [3]. Research polyheteroarylenes LC status is of great scientific and practical importance, as it allows to establish the general laws of phase transitions in polymers and to use the transition phenomenon in the mesophase to obtain high-strength fibers and films. The aim of this work was to study the temperature and concentration conditions for transition PABI H_2SO_4 solutions.

2 Experimental

In order to measure the temperatures of the liquid crystal $\rightleftarrows$ isotropic liquid solutions were placed in a sealed cell and asked by a teflon gasket layer of uniform thickness of 0,2 mm between the glass of the solution. The cuvette was then heated to the complete disappearance of the optical anisotropy and held under these conditions for 10-30 minutes. Control of the process was carried out to achieve a minimum amount of light transmission in crossed polaroid and visually – observation of dark field microscope in crossed polaroid (picture blanking) and the complete disappearance of H_v-scattering patterns. This temperature is taken for the transition to an isotropic (I) phase. The difference thus determined temperature values through a variety of polarization–optical methods was no more than 1-2°. The beginning of the reverse transition I → LC and the temperature is lowered is the appearance of the light transmission in crossed polaroid. We investigated the solution cooled to 20 °C and aged after cooling for a time sufficient to complete stabilization of light transmission. It is possible to eliminate the distortions introduced into the polarization-optical studies of stress anisotropy in a high concentrated solution and determine the critical concentration of the transition in the liquid crystal state. To eliminate the influence of crystal solvates the optical properties of the system and the study of phase transformations associated with the formation of LC phases namely, using the fact that the occurrence of crystal solvates due to the large induction period (from several hours to 2–3 days).

3 Results and discussion

The structural formula of this polymer is as follows:

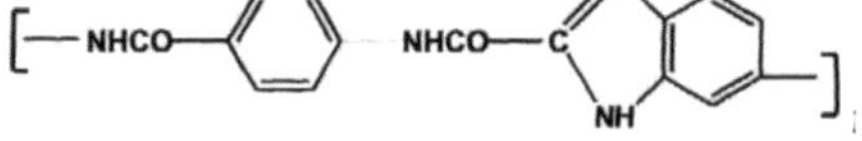

It was found that at concentrations in excess of 11% solution in H_2SO_4 PABI transformed into anisotropic state, similarly to forming solutions of PPTA typical lyotropic liquid crystal. PABI anisotropic solutions in 98% H_2SO_4 reprecipitated polymer was prepared from an intrinsic viscosity [η] = 5,8 dl/g measured in dimethylacetamide at 20 ° C. The solutions have a high viscosity (over 10^6 poise) were opaque and exhibited deformation characteristic "whitish", which disappeared after a while. Externally, the sulfuric acid solution with the concentration changed color from brown to whitish. In a sealed cell solutions show an interference pattern

* Corresponding author: o_hanchich@mti.edu.ru

 DOI: 10.1051/matecconf/20179901003

suppression, which disappears on heating and reappears on cooling.

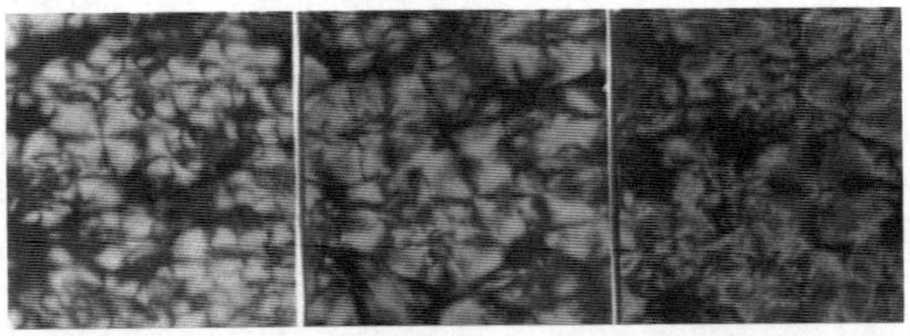

Fig. 1. The micrographs in crossed nicols of LC solutions of different concentrations PABI

Light scattering from such a sample has the form of a centrally symmetric spots, similar to the solution of the PPTA unoriented LC, with the intensity of the same order in H_v and V_v polarizations (Fig. 2, H_v, and V_v,a) that indicates the predominance of orientation fluctuations in the system and is typical of liquid crystal.

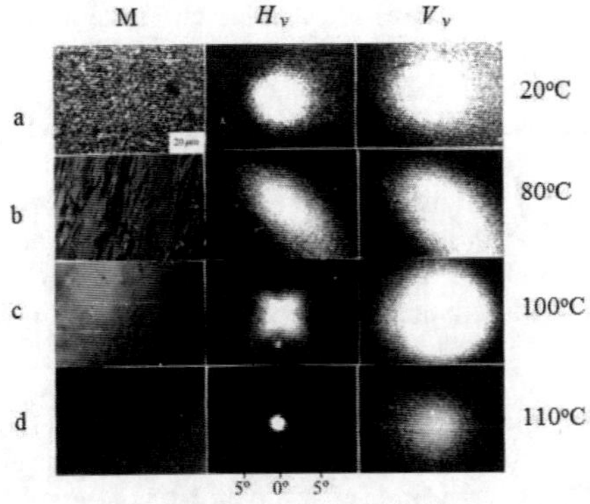

Fig. 2. The kinetics of structural change 15% – tion solutions PABI. M – microscopy, H_v – light scattering with crossed, V_v. – when parallel polaroids

The shear deformation of the solution leads to the transformation of the circular pattern of the scattering and the emergence of an elliptical diffraction patterns corresponding to the predominant orientation of cylindrical domains deformation direction (Fig. 2, b). Precipitation-based solution in water is accompanied by a decrease in the size of the ellipse and the formation of the equatorial scattering reflexes at high angles, which are located along the direction of the initial deformation of the solution appearance of such diffraction patterns can be explained by analogy with those observed for the PPTA. Increased shear strain λ_s half is split LC phase domains into smaller anisometric structure is perpendicular to the direction of shear. Subsequent precipitation of water does not lead to a qualitative change in the diffraction patterns, in addition to a displacement of reflexes in the region of small scattering angles, which corresponds to an increase in the size of the anisotropic structures. Analysis of diffraction patterns shows that all the basic laws of structure found in the deposition of anisotropic PPTA solution, hold for anisotropic solutions PABI in H_2SO_4.

It is of interest to determine the boundary of the LC status in PABI system – H_2SO_4 on polymer concentration and temperature, as well as study the kinetics of the system's transition to the isotropic state when heated and the emergence of an anisotropic phase when cooled. Studies were carried out on solutions with a concentration of 10-20%. In this case we use the full range of experimental techniques allowing the study of the temperature dependence of the scattering and extinction of paintings, as well as the overall intensity of the transmitted and scattered sample of polarized light. To select the optimal temperature conditions into account recommendations from the study of phase transitions in anisotropic solutions the PPTA. Phase transitions of the samples of anisotropic PABI (A) in the isotropic (I) state and vice versa, occur with certain combinations of the values of the critical concentrations and temperatures. It is possible to determine the status of the LC on the phase diagram of the system (Fig. 3).

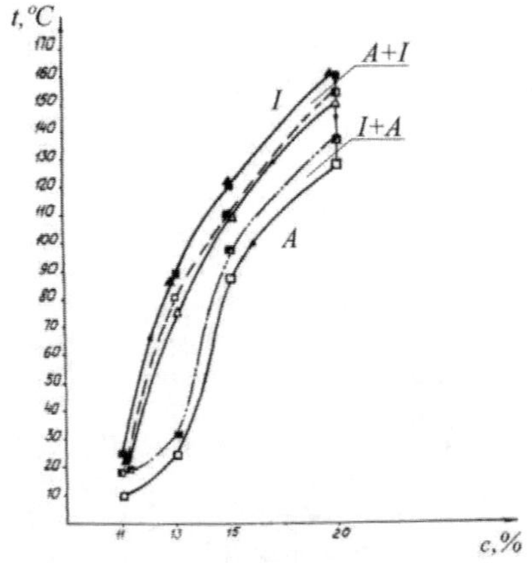

Fig. 3. Planning phase diagram PABI – H_2SO_4: ■ – the transition from the LC to the isotropic state (I) by heating, registered microscopy, □ – on the diffraction of light; ▲ – appearance anisotropic phase (A) under cooling registrant microscopy, Δ – diffraction light.

Used for plotting points disappearance anisotropic phase when heated, and its appearance upon cooling by microscopy and recorded by light diffraction. The sample temperature is changed near the transition region at a rate of 0,5 dg/min. As seen in Fig. 3, liquid crystal ↔ isotropic liquid (LC ↔ IL) phase transitions during heating and cooling are carried out for each of the combined concentrations in a certain temperature range The intensity of the transmitted polarized light to 11% solutions decreases monotonically with increasing temperature (Figure 4, curve 1). For the 13% solution is observed the appearance of a maximum intensity at a temperature of 63 °C (Fig. 4, curve 2), and for 15 and 20% - tion solutions - two peaks (Fig. 4, curve 3,4). The appearance of a maximum at lower temperatures can be attributed to the melting of crystal solvates. However, it must be borne in mind that extreme dependence of transmittance intensity can be caused by heating the mesophase transitions LC solutions.

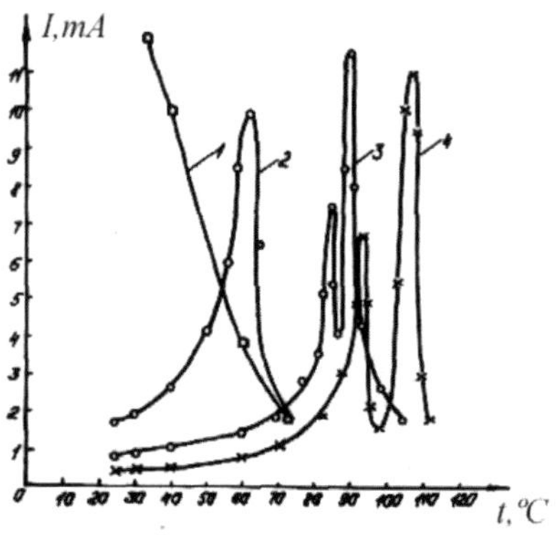

Fig. 4. Temperature the intensity of polarized light passing PABI sulfuric acid solutions of different concentrations,% 11 (1) 13 (2) 15 (3), 20 (4).

To answer this question let us analyze the kinetics of structural changes in the phase transition LC ↔ IL (Fig. 2). Heating LC sulfuric acid solutions PABI accompanied by a change in the color of the drug crossed nicols microscope (Fig. 2, M). Bright color is preserved up to 80 °C (a) 15% -s' anisotropic solutions. Separation occurs at 80 °C on a yellow color, red, green strip (Fig. 2, M, b). Stained bands varies with rotation of the microscope stage, which indicates the absence of optical activity and typical of the nematic liquid crystals. An increase in temperature leads to a red color (100 °C) of the drug (M, c), which disappears completely (118 °C) at the transition to the isotropic state (Fig. 2, M, d). Upon cooling, the drug is an alternation of colors in reverse order, but at slightly lower temperatures. Such changes, but other temperatures observed during heating and cooling of the solutions in the other concentrations (11, 13, 16, 20%).

It is known that light scattering method in conjunction with other techniques allows to classify the type of liquid crystal and to study changes in the structure [4]. By virtue of the specific interaction with the molecule of H_2SO_4 PABI arranged in a certain order to form domains that are commensurate with the length of the incident light. Reduction of the scattering angle θ indicates an increase in domain size and appearance of the scattering intensity depending on the azimuthal angle μ may be associated with a transition to a different texture LC.

Upon heating the solution several angle θ increases, the size of which corresponds to a reduction of ordered regions of liquid crystal. Existing in a certain temperature range anisometric structure (ribbon-shaped inclusions of different colors visible, and light scattering (Fig. 2, M, b). At these temperatures, the diffraction patterns become elliptical form (Fig. H_v, V_v, b). Changes in the nature of the orientation of the structural units from a disordered to an ordered, manifested in the transition from circular to elliptical scattering pattern. This may be due, in analogy with low molecular weight liquid crystal [5], the existence of a temperature gradient between the cell substrates with anisotropic PABI solution, which leads to the preferential orientation of the molecules. Areas of anisotropic and isotropic phases coexistence correspond significant changes in the intensity and shape of the scattering of paintings (Fig. 2, H_v, V_v, b). Near the transition to the isotropic phase the intensity of the diffraction pattern becomes dependent on the azimuth angle μ (Fig. 2, H_v, V_v, c). This intensity distribution can be called molecular groups in a cholesteric state [3].

Heating and cooling rate not only change the temperature range of existence of phases, but also affects the structure of the sample. For example, the rapid cooling rate of 10 °C/min, 15% solution, heated to 20 – 30 °C above the transition temperature to the isotropic state, leads to the formation of the anisotropic phase in the form of diamond-shaped inclusions.

Loss LC embryos observed in a similar form of low molecular weight smectic liquid crystals [6, p.11]. However, based on this phenomenon can not assert a priori about the transition to the smectic mesophase, since when heated to these temperatures degradation of the polymer is also possible, which may lead to the formation of structures of this type on crystallization of degradation products. On the validity of the latter indicates a sharp decrease in the intrinsic viscosity of the system. It was also found a difference in the structure of anisotropic solutions of different concentrations PABI. 20% solutions formed in a temperature range of symmetric H_v scattering pattern with peaks at an angle μ = 45° to the direction of polarization. Diffraction patterns with azimuthal dependence of a different type (μ = 0 and 90°) were obtained upon cooling 15% solutions. As noted, a similar distribution of scattered light intensity in the direction corresponding to scattering of light from bar-like structures with main polarizability axis oriented at an angle of 45° to the axis of the rod.

Observed by optical methods polymorphic transformations are also recorded on DTA thermograms (Fig. 5).

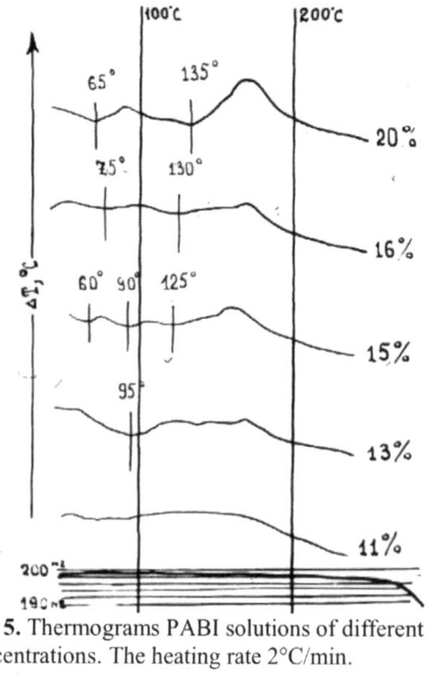

Fig. 5. Thermograms PABI solutions of different concentrations. The heating rate 2°C/min.

The cycle of heating (cooling) LC PABI solutions of different concentrations are observed in the thermograms small in size (1 cal/gram) endoeffects showing the restructuring of the liquid crystal by heating. Maximum peak (5 cal/g) endothermic region with decreasing concentration of the solution is displaced by 140 °C (20% solution) to 95 °C (13% strength) and not accompanied by changes in sample weights. This shift corresponds to the transition LC ↔ IL by heating the solution.

4 Conclusions

By using polarization-optical methods and the structural features of biphasic and anisotropic areas and built plots the phase diagram of the concentrated solutions PABI in H_2SO_4. It is found that to obtain an LC-based systems can be directly PABI solution PABI H_2SO_4 at a concentration above 11%.Thermogram, as well as optical findings indicate that structural changes LC PABI solutions occur in a temperature range which increases with increasing concentration of polymer in solution. Under the influence of the temperature in the system may be structural changes associated with melting of the crystal solvates, as well as transitions from one to another modification of the LC, similar to that observed in [7]. Use of H_2SO_4 as a solvent alters not only the temperature - concentration phase transitions, but also the structure of anisotropic solutions PABI. In the transition PABI solutions in the anisotropic state during cooling are formed spherulites. Wherein in certain temperature - concentration of cooling modes can be observed the coexistence of three phases: isotropic crystal and a liquid crystal, which is shown as a characteristic of liquid crystal birefringent domains. As already mentioned, there are crystal solvates [8] at high concentrations of polymer solutions in sulfuric acid [8]. The regularities enable the correct selection of the conditions of implementation of the mesophase in other practically important polymers.

References

1. K.E. Perepelkin, *The reinforcing fibers and fibrous polymer composites* (Moscow, Publishing House of Scientific basis and technology, 2009)
2. SP. Papkov, VG Kulichikhin, *The liquid crystal polymer state* (Moscow, Chemistry, 1977)
3. O.A. Khanchich, S.A. Kuznetsova, *Cloud of Science*, **1**, 30-32 (2013)
4. O.A. Khanchich, *Anisotropic structure in polymers with varying rigidity of macromolecules* (Thesis of Doct. Chem. Sciences, Moscow, 1997).
5. P.G. De Gennes, *The Physics of Liquid Crystals* (Mosocw, Mir, 1977)
6. I.G. Chistyakov, *Liquid crystals* (Moscow, Nauka, 1966)
7. E.M. Antipov, S.D. Artamonov, I.A. Volegova, J.K. Godowsky, *Vysokomolek. Comm. A*, **37**(5), 800-810 (1995)
8. V.G. Kulichikhin, N.V. Vasileva, V.A. Platonov et al., *Vysokomol. Comm. A*, **18**(3), 590-596 (1976)

Non-destructive Assessment of Relief Marking Parameters of Heat Shrinkable Installation Parts for Aviation Technology

Aleksandr P. Kondratov[1, *], *Mariya V.* Konovalova[1], *Egor P.* Cherkasov[1] and *Irina A.* Savenkova[2]

[1]Moscow Polytechnic University, 107023 Moscow, Russia
[2]National Research University Higher School of Economics, 101000 Moscow, Russia

Abstract. The article explains a new method of relief marking of heat-shrinkable tubing and sleeves made of polymer materials with "shape memory effect." Method of instrument evaluation of relief marking stereometry of installation parts for aviation equipment, made of polyvinyl chloride, polyethyleneterephthalate and polystyrene was developed and the results were explained. Parameters of pin-point relief marking and compliance of point forms to the Braille font standard were determined with the use of the non-destructive method based on the color of interference pattern with precision of 0.02 mm.

1 Introduction

For new samples of complex machinery [1] in aviation industry and aircraft-repair works [2, 3], polymer heat-shrinking tubes and sleeves are used with «shape memory» in order to provide additional mechanical protection, sealing action and electrical insulation of contact, complex connections of electrical wires etc.

Various types of marking are used in order to control technical conditions of various aviation components and determining the appurtenance of communication tools. As a rule, multicolored isolation cable sheaths, tags and flags are used for the purpose of identification. In the aircraft electrical systems all wires, harnesses and cables are subject to marking [4]. Appurtenance of wires to various systems and feeders is determined not only by wire insulation color and by marking, that presents an alphanumeric code, which is put on polymer insulation of the wire or on its identification tag. In the absence of direct access, insufficient visibility or bad lighting conditions of the communication location it is feasible to use relief marking.

In order to make the process of reading the printed information on the insulating jacket or sleeve, that is, for example, carried out with the usage of Braille, relief symbols should have a specific appearance, specific location on the plane and comply with strictly specified stereometry parameters [5].

The technology of applying the relief tags on the products made of heat-shrinking materials [6, 7] that was developed by us allows to record information on soft and flexible polymer films with compliance of allowable sizes of Braille font symbols. In order to determine the optimum conditions and control the process of product tagging (labels, packaging, couplings and tubes) made from polymer materials, instrumental quantitative evaluation of micro-size relief stereometry is required.

There are a few methods of analyzing the dimensional transparent microscopic objects which are used in medicine, forensics, materials science [8, 9].

In previous experimental works [10, 11] measurement of parameters of tactile relief was carried out in the following way. Sample of elastic label product was cooled up to the temperature of below the polymer glass-formation in order to increase the rigidity and was dissected with curring blade along the relief crest. Afterwards the received cut-section was photographed dimensions of the image were measured and evaluated, with the usage of gauge-metering function of optical digital USB-microscope. At the stage of sample dissection deformation of the tagg's relief is unavoidable, consequence of which is the measurement error in the amount of + 0,3 micron.

2 Goals and objectives

Development of a high precision instrumental procedure of non-contacting evaluation of stereometry of relief marking of flexible and elastic polymer parts with «shape memory», that are applied during installation and repair of aviation unit communications.

3 Materials and methods

Heat-shrink tubing and polymer films of industrial production from polyvinyl chloride, polyethylene terephthalate and polystyrene, which thickness is 50 microns. Producers of heat-shrinkable materials are Dongil Chemical and Alfterm. Coefficient of maximum shrinkage during T = 90±5°C heating should be around 70%. Relief point-like and linear signs should be applied on the samples of polymer films. Point-like tagging was carried out with the usage of soldering station that provides temperature control at 100±1°C of cylindrical

* Corresponding author: apkrezerv@mail.ru

soldering iron rod with a flat foundation and diameter of 1.0 and 3.0 mm. Linear tagging was carried out with the usage of impulse heating unit with local thermal processing [12]. «Developing» of relief was carried out with settlement on holding base (mounting jig) that imitated the isolated communication line of aviation unit through heating with a hot air flow or in water (85-95°C) by the special method [10].

4 Results and discussion

Algorithm of the new method of relief marking of heat shrinkable parts made of thermoplastic polymers includes the following operations:

Local thermal processing of samples (heat-shrinkable tubing or film) in isometric constraints, application of the sample on a rigid base (communication tools, wiring plugs) and heat shrinking process in some hot heat-transfer medium (air, water).

Local thermal processing of the product with «shape memory» is carried out in isometric conditions. During local thermal processing of the small product section the isometric mode is provided by the hardness of the polymer material that surrounds the area of local polymer heating. As a result of hardness of glass-like polymer the shrinkage of the heat-treated part doesn't take place, instead local relaxation process of internal stresses happens. Stabilization of geometrical dimensions of the local thermal processing area is a result of relaxation of the internal stresses, which means the loss of ability for heat-shrinking.

During heating of the product after local heat-treatment process through submerging the sample into the hot medium its surface drastically reduces. Areas of local thermal processing do not change their dimensions of thermostabilized area and thus it bulges out. Relief is formed, that contains visual and tactile information. Relief can have the form of vectors or geometric shapes or form of dots with the stereometry of Braille font. Main factors that determine the form and dimensions of the relief are shrinking properties of the polymer (coefficient of thermal shrinkage), dimensions of the areas of local thermal processing (diameter of the dot foundation or width of lines) or ratio of heat-shrinkable tube's (sleeve) dimensions and dimensions of communication tools where the enclosure or sleeve are applied.

Types of relief labeling on heat-shrinkable pipes that are applied in electrical and electronic communication tools of aviation units are shown as a scheme on the Fig. 1 and 2. Non-contact composition of 3D models of relief of the experimental sample was used for instrument evaluation of stereometry of aviation equipment installation parts surfaces made of transparent polymer materials. Laboratory direct microscope of reflected light Carl Zeiss Axio Lab.A1, equipped with additional measuring device PhaseView, was used for this purpose. Determination of the distance to the fragment of the surface was carried out through analyzing the interference pattern that is formed inside the additional measuring device. This device contains calibrated emitting source, which allows measuring the precise dimensions of the relief with use of results of preliminary measurement of the distance to the researched sample based on location of maximum points and minimum points of interference pattern. Geometric data of the optical path that are contained in calibration files of the used software are used for preliminary measurement of the distance.

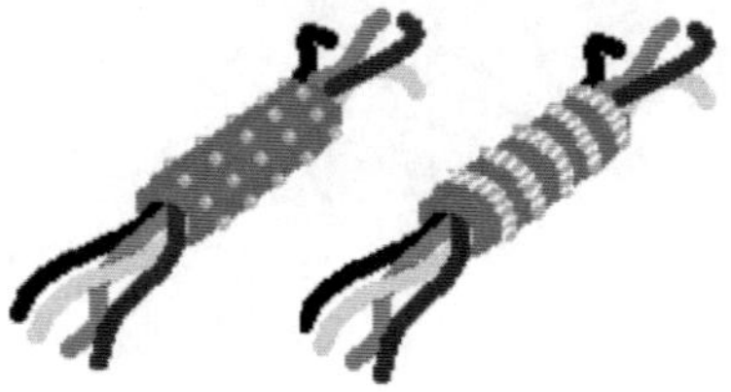

a b

Fig. 1. Type of sections with the relief labeling of the products with «shape memory» from heat shrinkable shelter of multicore cable: a) pinpoint marking; b) linear marking

Fig. 2. Types of section with relief labeling of separate wires made from heat shrinkable tubes

In order to carry out the optical tests we have to cover the surface of the polymer tube sample or film with pigment or dye solution of any dark color, because of distortion-noises, which occur on the projections of transparent objects. Coloration of the surfaces of researched objects was carried out with the use of inks that contain ethanol.

Building of 3D model is provided by GetPhase software. The relief of the surface looks much more informative in isometrics (Fig. 3). The height of characteristic points of the surface relief is based on the color tone range (shown on the right on the Fig. 3a, b).

Quantitative characteristics of the relief that were received during measurement of the relief marking of elements of various types are given in the Table 1 and Table 2. Geometry of the marking [13] is not very varied, it can either be pin-point or linear, but the dimensions of the marking can change depending on the area of processed polymer surface.

Table 1 contains the information of overall dimensions of pin-point marking of polymers with different chemical composition, but with the same thickness. From the table it is clear, that the more a diameter of the heated element is, the higher each point becomes. Table 2 presents results of the amplitude and period measurements of the relief marking. As a result, because of the fact that polyethylenterephtalate has higher hardness, the relief that was formed by pin-point and linear marking on these films samples is quite high in comparison with relief on other films, used in the experiment.

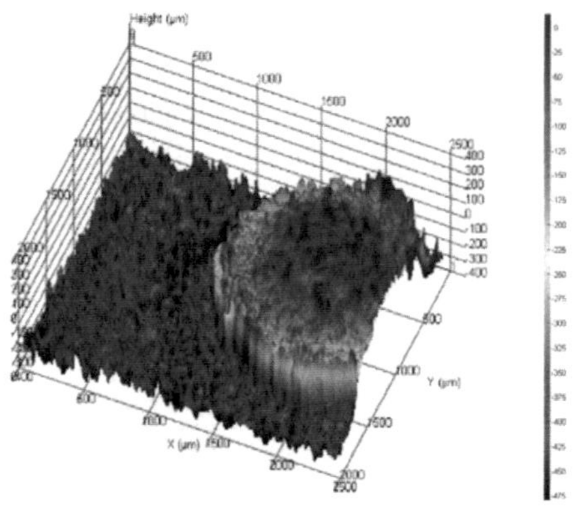

a

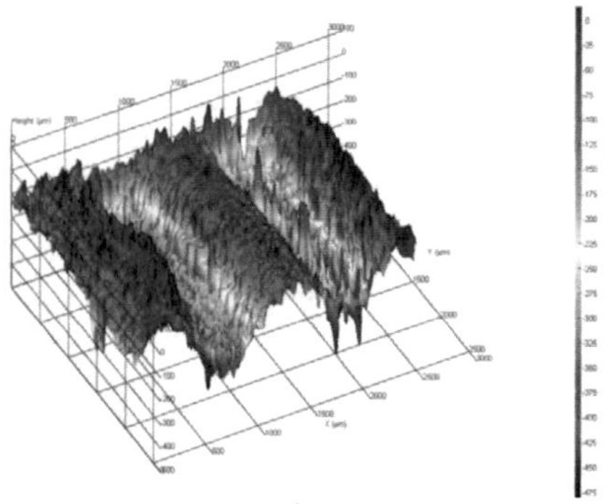

b

Fig 3. The 3D-Model of fragments of various relief marking of polymer films: a) Pinpoint marking; b) Linear marking

Table 1. Parameters of pin-point marking of the heat-shrinkable polymer samples

Polymer material	Height of the point, μm	
	Diameter of heated element -3.0 mm	Diameter of heated element -1.0 mm
Polyvinyl chloride	442	370
Polystyrene	563	243
Polyethyleneterephthalate	667	30

Table 2. Parameters of the linear marking of the polymer film with the thickness of 50 μm

Polymer material	Parameters	
	Amplitude, μm	Period, μm
Polyvinyl chloride	264	1180
Polystyrene	462	1660
Polyethyleneterephthalate	512	1720

5 Conclusion

The quality of visual control and preventive measures against shutdowns of electrical networks in aviation equipment very much depends on the quick identification of communication tools (wires, cables, tags and flags). Their identification is based on the color differences or quick finding and reading the printed markings, what in conditions of bad visibility and limited access to the units of aviation equipment can be quite challenging. Thus, the presence of relief symbols on heat shrinkable tubes will considerably simplify the process of identifying the communication linkages.

Additionally, the developed instrumental method of the non-contact evaluation of stereometry of relief marking of felxible and elastic polymers allows carrying out more detailed studying of the investigated samples' surfaces. Without breaking the integrity of the sample, measurements were carried out with high precision and high accuracy.

References

1. V.A. Kaniskin, *Operation of power electric cables: Teaching guide* (Saint-Petersburg, PEIPK, 2001)
2. N.M. Ilyshin, *Methodical recommendations on organizational maintenance and first line repair of electrical systems of aircraft vehicles* (Lyubertsy: 13 GNII RF Ministry of Defence, 1985)
3. Airplane Il-76 /Battle Damage Repair Manuals/ Section Aircraft Electrical System (LLC "Avia-Media", 2004–2007)
4. Braznburg, E.Z. et al. *Cables iwht plastic insulation and couplings for their installation* (Moscow, Energoatomizdat, 1987)
5. Marburg Medium spacing convention for braille. Website of the company Deutsche Blindenstudienanstalt e. V Braille-Druckere Postbox 11 60 35001 Marburg : http://www.blista.de/download/druckerei/braille-dimensions.
6. M.V. Konovalova, A.P. Kondratov, The patent of the Russian Federation № 150223 (2015)
7. A.P. Kondratov, *Izvestia vuzov. Problem of the graphic arts and publishing*, **4**, 57-65 (2010)
8. Methods of research in histology, cytology and embryology. http://profmed.blogspot.ru/2013/04/blog-post_28.html. Date of access 03.03.2016
9. V.B. Konstantinov, V.A. Babenko, *Applied physics journal*, 77 (2007)
10. M.V. Konovalova, A.P. Kondratov, *Izvestia vuzov. Problem of the graphic arts and publishing*, **3**, 53-59 (2015)
11. A.P. Kondratov, *Modern Applied Science*; **8**(6), 204-209 (2015)
12. M.V. Konovalova, A.P. Kondratov, V.G. Nazarov, A.F. Benda, *46th Annual International Conference on Graphic Arts and Media Technology Managem Management and Education, Athens and Corinthia, Greece*, p. 251 (2014)
13. E.P. Cherkasov, A.P. Kondratov *Izvestia vuzov. Problem of the graphic arts and publishing*, **1**, 43-47 (2015)

Geometric Simulation of Design Objects in Aerospace industry

Alexander Chursin[1], *Alexander* Semenov[1, 2, *] and *Anna* Ostrovskaya[1]

[1]People's Friendship University of Russia, 117198 Moscow, Russia
[2]Moscow Technological Institute, 119334 Moscow, Russia

Abstract. The article outlines the approach used for geometric simulations of design objects in aerospace industry. The computational model describing the effect is presented, which is recommended for practical use in construction.

1 Introduction

While designing a new item the designer is aimed at the development of functions and geometry, then it is required to arrange these results in specifications. Below we will outline the basic approaches which we recommend to use in aerospace industry (basic approach was developed in [1]). The importance of new methods and innovations in aerospace industry is described in [2].

Geometry defines shape and dimensions of an item and its components, its functions describe operational principles and interaction of its subcomponents. Specifications are supplemented by information about the material and processing techniques.

In the course of designing the designer not only develops concepts of item's geometry but works with it using established practices of geometry description. According to operating techniques, up till now in designing bureaus the designer expresses concept of the item's shape and dimensions in the form of 2D presentation – engineering drawing.

Computer aided design can be promoted by engineering implementation and computerized presentation of the mentioned concepts of the item, as well as possibility of their processing at various levels of abstract description.

The term "specifications" involves all information required for the item production which should be presented in the form, oriented at certain production technique. This concept covers all engineering drawings, specifications and flowcharts. Technical specifications embody the most important information required for the produced item. Computerized development of concepts of the item and computerized processing of the information in specifications required for relevant description.

Geometric object requires description of the item using mathematical model in Cartesian space, and the presentation is arranged with consideration of the shape and dimensions. Presentation of functional links in the item, required upon designing, is not taken into account at this stage. The mentioned geometric objects are mathematical (algebraic) structures presenting more or less accurate embodiment of the item's geometry in the mentioned aspect. The presentation and description of such geometric objects are based on analytical geometry.

2 Geometric simulations

Problematic procedures, which should be applied to the item in the frames of considered engineering tasks, are referred to the item description as geometric object. Interrelation between such abstract geometric object with the array of human concepts is performed by presenting geometric object in the form of graphic presentations.

A graphic presentation means projection of 3D geometric object onto 2D plane. A particular case includes 2D geometric objects.

Interrelation between geometric object and human reasoning, together with object graphic presentation, can be performed by fabricating the material model of a geometric object.

Specifications and graphic presentation are closely related by dependency relation, because graphic presentation is a specialized element of specification.

Selection of reasonable form for describing geometric objects and graphic presentations is defined by actual practices and type of processed items. One of the classifications, oriented at industrial items, stipulates the following classes of geometric objects: prismatic, revolved bodies and arbitrary shapes. Prismatic items are restricted by planes; revolved bodies, respectively, by the second order surfaces (cylinders, cones, spheres); and items of arbitrary shapes by surfaces of higher orders or surfaces which cannot be described exactly. The task is set in principle as follows: in terms of analytical geometry the initial elements can be defined as points, lines, curves, planes, surfaces, as well as bodies. This set of objects is used to generate new objects

* Corresponding author: semenov.venture@mail.ru

according to the established syntax. The generation is performed as the recursive procedure, that is, the objects obtained from the main objects can be combined with other main objects into new objects. Two contradictory requirements are related with such description of geometric object:

– the description should be suitable for subsequent processing using computer aided designing;

– the description should be structured to such extent than it could be used for development of graphic presentation, that is, sequence of instructions controlling motions of the plotting instrument or the electron beam.

Flat objects are described by the so called simplest elements: primitives. The primitives are considered as 2D objects characterized by two highlighted points: the beginning and the end. These points are the sites of possible combination of the primitives. The simplest elements are described in terms of classes containing the elements' names.

The names facilitate recognition of data for initial and final points, as well as determination of other element's characteristics. This description is applied for development of drawings, graphic presentation of details on flat screen. Analysis of numerous engineering drawings demonstrates that about 99% of all drawing elements are segments of lines and circular arcs.

3D geometric objects are described on the basis of the following algebraic structures:

G - graph structure;

F - surface element structure;

K - body structure.

G – structure, that is the object comprised of points on plane or in space $P(p_i)$, where P is the main set of all allowed points p_i. Merging and attachment operations are allowed. The merging operation designated by " - " symbol corresponds to the segment of the line, connecting points $\boldsymbol{p}_1$ and $\boldsymbol{p}_2$: $(p_1 - p_2) = x$. The obtained element belongs to the G-structure. The initial point of oriented element $\boldsymbol{x}$ is $\boldsymbol{S}(\boldsymbol{x})$, the final point is $\boldsymbol{l}(\boldsymbol{x})$. The attachment operation is designated by " . " symbol. It interrelates the points $\boldsymbol{p}_1$ and $\boldsymbol{p}_2$ and attaches them with G-structure without generating an additional element $(p_1 \text{ and } p_2)$.

Selecting " $\circ$ " as the sign for " - " operations in " . ", taking $x = g_1 \circ g_2 \circ g_3 \circ g_m$, it is possible to describe any structure plotted from nodes $\boldsymbol{P}$ and edges $\boldsymbol{K}$, where $\boldsymbol{K} = K\{(p_1 - p_2)\}m_{ij}$ is the set of graph's edges; G is the set of points and segments and $<$ **G**, - , .$>$ is a certain algebra.

This form of representation (point model) can be considered as geometric image of a body restricted by flat surfaces. Graphic images of such geometric object are the so called transparent or wire lattice models (polyhedra).

Graphic images of polyhedra are presented by flat graphs. The presented edges are intersection lines of surfaces restricting the geometric object. The items are presented in the form of transparent wire frame.

F-structure serves for presentation of bodies and surfaces composed of elements. The main elements are curvilinear tetragons, described parametrically: $X = X(U,v)$, $y = y(U,v)$, $Z = Z(U,v)$. Paramtrized element is defined by logical expression:

$$\boldsymbol{e} = (\boldsymbol{X} = \boldsymbol{X}(\boldsymbol{U}, \boldsymbol{v})) \wedge (\boldsymbol{y}(\boldsymbol{U}, \boldsymbol{v})) \wedge (\boldsymbol{Z} = \boldsymbol{Z}(\boldsymbol{U}, \boldsymbol{v})) \wedge$$
$$\wedge (\boldsymbol{b} = \boldsymbol{b}(\boldsymbol{U}, \boldsymbol{v}))$$

which equals to $\boldsymbol{I}$, if the point $p(x,y,z)$ belongs to the surface element. Double valued function $b(\boldsymbol{U},\boldsymbol{v})$ serves for element's restriction. It equals to $\boldsymbol{I}$ inside of the element and $\boldsymbol{O}$ outside of it. In order to integrate two point sets, two surface elements are merged by disjunction $f = f_1 \vee f_2$. Here $(f, \boldsymbol{v})$ can be considered as an algebra.

K-structure is intended for plotting the bodies using logical functions of disjunction $K = K_1 \vee K_2$ and conjunction $K = K_1 \wedge K_2$ similar to merging of point sets, since together with plotting the bodies using the principle of indication of surface restricting elements, it is necessary to generate bodies directly of main (basic) bodies. The main objects (bodies) are the functions $\boldsymbol{a}(x,y,z)$ taking real values. For each expression $\boldsymbol{a}$ it is possible to determine:

main closed body $h(x,y,x) = [\boldsymbol{a}(x,y,z) < 0]$;

main open body $h(x,y,x) = [\boldsymbol{a}(x,y,z) > 0]$;

main surface $h(x,y,x) = [a(x,y,z) = 0]$.

Here $K, \vee, \wedge, >$ can be considered as algebra.

3 Conclusions

Descriptions of hierarchical generation of geometric objects and graphic images depend on a specific setting of tasks, thus, it is actually possible to obtain various modifications as well as particular cases of the aforementioned forms of description.

References

1. V. V. Efremov, V. A. Naumov, A. A. Chursin, *Theory and practical issues of operability of elements of machines, devices and instruments*, (Irkutsk University, Irkutsk, 1984) [In Rus]
2. A. A. Chursin, A. S. Semenov, A. V. Danilchankau, NPCS **19**, 3 (2016)

The Synthesis of Electric Drives Characteristics of the UAV of "Convertiplane–Tricopter" Type

O. V. Emelyanova[1, *], *G. K.* Kazaryan[1], *A. S.* Martinez Leon[1], and *S. F.* Jatsun [1]

[1]Southwest State University, 305040 Kursk, Russia

Abstract. The paper is devoted to the review of controlled motion of the unmanned aerial vehicle (UAV) with vertical takeoff and landing, the convertiplane type, with three tilt rotors. The UAV is equipped with a swept wing, the calculation of aerodynamic parameters of which has been carried out in the software program XFLR5, taking into consideration the weight and size requirements. There have been determined the lift and resistance coefficients of the UAV depending on the angle of attack on the basis of the vortex method. There has been done the analysis of the influence of wing profile on the calculated aerodynamic characteristics. There also has been realized the synthesis of the traction energy-efficient options of the drive propellers that reduce power consumption, increasing the accuracy of the automatic control system.

1 Introduction

Currently, the UAVs of various structural and constructive models, performing vertical takeoff and landing, are actively used in many spheres of human activity. It is connected with the fast growing technology and engineering solutions. As the result, such devices have become compact and maneuverable, simple in production and controlling, with a reliable and lightweight construction that has been achieved through the use of modern composite materials, providing a low weight of these devices with a significant payload mass [1–4, 5, 11].

Convertiplane can be considered as one of the most promising types of UAVs, because it combines helicopter features – a vertical takeoff (landing), and aircraft features – a flight in a horizontal plane if there is a wing, by changing the traction vector of propellers [5–8, 13–15]. The advantage of such device is the energy consumption reduction of the on-board system that can be achieved by turning the traction forces of the propellers in the direction of movement, and using the flying wing, as the significant part of the aerodynamic force is formed upon it. The convertiplane flight can be divided into several stages: the climb and decrease on the vertical trajectory, and the flight in the horizontal plane. The modes of vertical takeoff or landing are more complex than a horizontal flight, as there is an additional resistance of the fuselage to the wing. In addition, the fuselage of the convertiplane has a bad streamlining during the vertical rise. Therefore, in these modes, the traction of the lifting propellers increases substantially. [1–4].

The horizontal flight of a convertiplane requires less traction, because in this case the fuselage and the wings are streamlined by air and their frontal resistance decreases. The resistance of the lifting propellers also reduces as a result of their transition to a smaller angles of attack or the turning in the direction of movement. At any speed of the convertiplane horizontal movement the lift of the wing allows to minimize the height loss at the moment of the propellers turning, providing the necessary stability of the flying vehicle. In this case an important role plays the calculation of the aerodynamic parameters of the wing, which can significantly increase the carrying capacity of the convertiplane during a horizontal flight.

2 Description and principle of movement of the convertiplane

Let's consider the motion of a convertiplane in a fixed cartesian coordinate system *OXYZ*, where $CX_1Y_1Z_1$, $A_ix_iy_iz_i (i = 1, 2, 3)$ are movable coordinate systems passing through the mass center of the device body *C* and the mass center of the *i* -rotor, respectively (Fig. 1). The space orientation is set by the plane yaw ψ, pitch θ and roll φ angles.

The device consists of three pairs of controlled propellers 1-3, on the basis of the brushless electric drive with variable traction vectors and the fuselage 4, on which there also fixed a power supply, a control board and a signal receiver that are electrically connected with the rotary drive of the propellers.

The declination of the rotary drive traction vectors 1, 2 is carried out in the plane $A_iX'_1Z'_1$ (|| CX_1Z_1) at an angle β, which may vary from 0 to 90^0, and the drive 3 – in the plane $A_3Y'_1Z'_1$ (|| CY_1Z_1) at an angle $\pm 45^0$.

* Corresponding author: teormeh@inbox.ru

The presence of the traction horizontal component of the tail drive will allow providing the rotation of the convertiplane around the vertical axis CZ_1 on the yaw angle ψ. The torque compensation takes place due to the oppositely installed propellers, turning in opposite directions [8, 9].

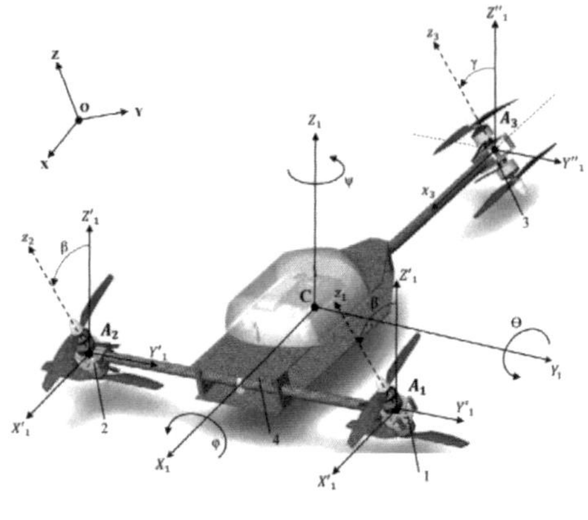

Fig. 1. The general view of a convertiplane-tricopter power frame.

3 Calculation of the convertiplane aerodynamic parameters

When calculating the basic convertiplane performance characteristics, as well as its stability and controllability, it is necessary to know the forces acting on the device during the flight. Conditionally, they can be divided into the following forces (Fig.2) [5–7]:

1) in absolute (Cartesian) coordinate system *OXYZ*:

- the gravity $\bar{G}$ of the convertiplane applied in the mass center *C*;

2) in a moving coordinate system $CX_1Y_1Z_1$:

- the forces, generated by the propellers $\bar{F}_1 - \bar{F}_3$, applied in the mass center of the rotors A_1–A_3;
- the aerodynamic forces, acting on the wing surface, called the force of frontal resistance $\bar{Q}_i$, directed to the opposite side of the axis CX_1, and the lift force – $\bar{P}_i$, directed along the axis CZ_1 (Fig. 2) and, similarly, the aerodynamic forces, acting on the fuselage: $\bar{Q}_\Phi$ and $\bar{P}_\Phi$.

The aerodynamic forces, acting on the wing surface, can be expressed via dimensionless coefficients of the frontal resistance CD and the lift CL. Due to the smallness of the side force coefficient CP, the aerodynamic side force $\bar{Q}$P, won't be taken into account in this modelling [3].

Then

$$\begin{bmatrix} Q \\ P \end{bmatrix} = q_\infty S \begin{bmatrix} C_D \\ C_L \end{bmatrix}. \qquad (1)$$

where $q_\infty = \rho v_\infty^2 / 2$ the dynamic velocity pressure; v_∞ – the air rate; S – the wing area.

The aerodynamic characteristics of the wing depend on the geometric parameters of the wing profile, the orientation of the wing in the airflow (angle of attack α), similarity parameters, etc. Therefore, for their determination the results of analytical, numerical and experimental studies, founded in XFLR5 software product under specified geometric and aerodynamic characteristics of the airfoil will be used.

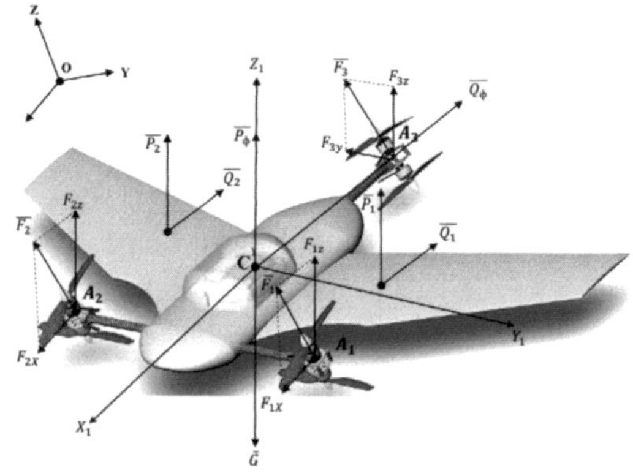

Fig. 2. The scheme of forces, acting on a convertiplane during the flight

The selection of the airfoil is one of the main quality indicators in the study of the UAV aerodynamic characteristics. For the flying vehicle under review the selection of the profiles has been made from the standard (normal) profile classes "FX" and "NACA", which, if necessary, can be modified at the stage of conceptual design (Fig. 3). This choice is determined by the fact, that the profiles of represented classes have low weight, acceptable thickness, the maximum time of floating in the air, the highest lifting capacity of all the standard series, the stability at low and high values of the number Re (laminar and turbulent flow) [13]. The profiles are asymmetric, because the pressure difference, occurring at the lower and upper edges, provides a greater value of the lift force than on the symmetrical ones.

Further, the effect of the selected wing profiles and the swept wing shape on the main vector of the aerodynamic forces will be considered. The choice of a swept wing shape is determined by the fact, that it has a high lift coefficient at small angles of attack [1,3]. This allows reducing its dimensions and weight. For modeling the shape, there have been used the following geometric parameters (Fig. 4): the root chord (b_0), the mean chord (b_{cp}), the tip chord (b_k), the sweep angle (α) and the wing length (l_{wing}).

Next, there will be determined the optimum ratio that provide the maximum value of the lift coefficient C_L from the angle of attack α for different types of profiles. The modelling has been carried out with the following parameters: the air density ρ = 1,225 kg/m^3, the speed of the oncoming air flow υ = 5 m/s. The modelling results are shown in Fig.5.

DOI: 10.1051/matecconf/20179902002

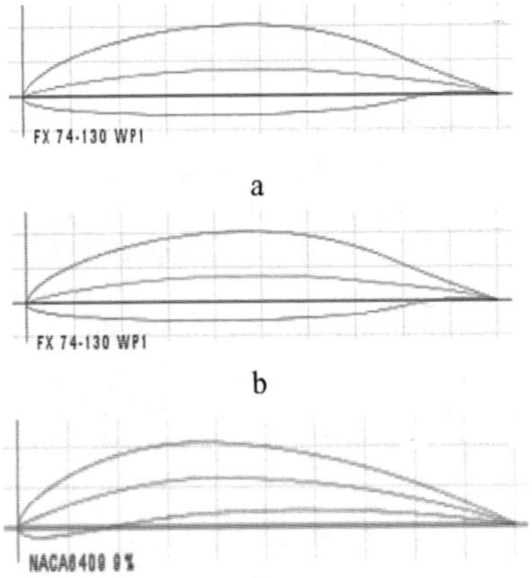

Fig. 3. The profiles under study: *a* – FX 74-130; b – FX74-CL5-140; *c* – NACA 6409 9%

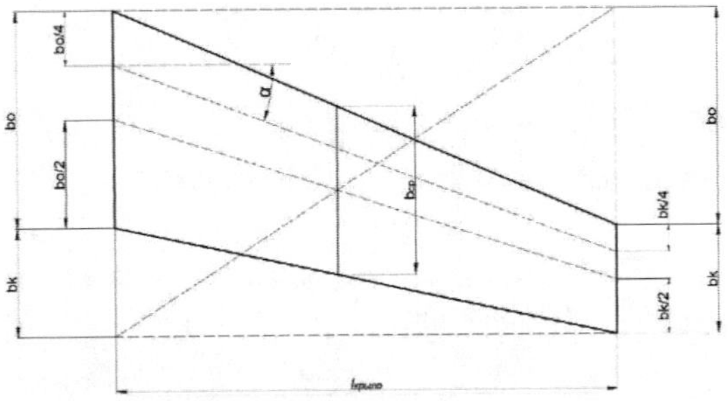

Fig. 4. The geometric parameters of a planned wing

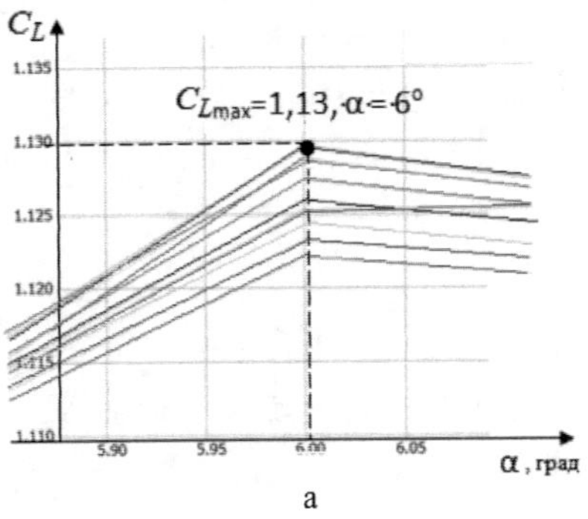

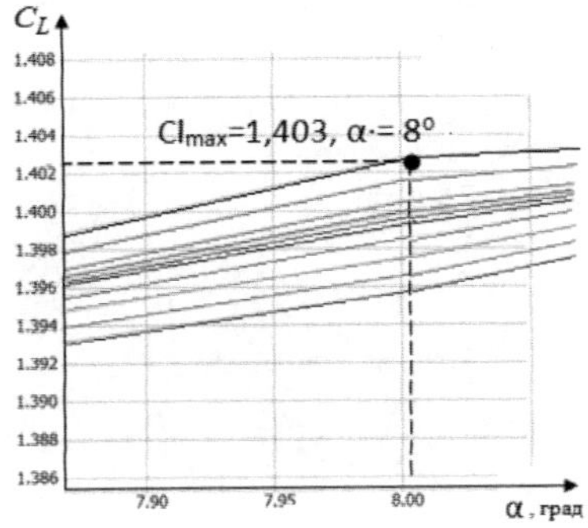

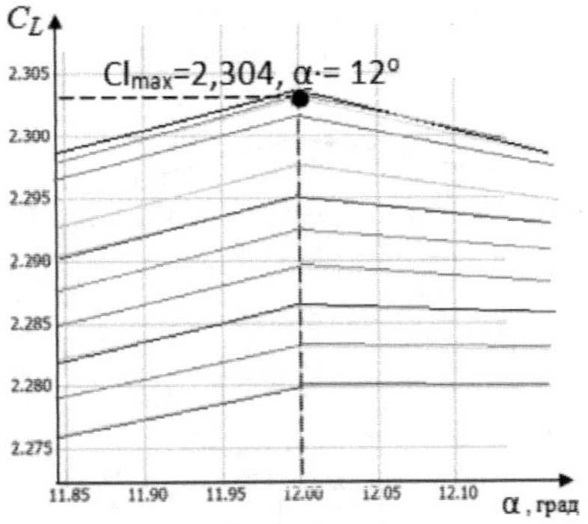

Fig. 5. The lift coefficient C_L – angle of attack α diagram, using the profiles: *a* – FX 74-130; *b* – FX74-CL5-140; *c* – NACA 6409

Similarly, there will be determined the optimum ratio, providing the maximum value of the resistance coefficient CD from the angle of attack α for different types of profiles.

Then, taking into consideration (1), there will be calculated the aerodynamic lift and frontal resistance forces for the swept wing shape and different profiles. The results of the study are presented in table 1.

Table 1. The aerodynamic parameters of the wing

Form	Profile	α [°]	Cl	Cd	S [m^2]	$\overline{P}$, [H]	$\overline{Q}$, [H]
	FX74-130	6	0,81	0,03	0,22	2,72	0,11
	FX75-CL5-140	8	1,44	0,10		4,83	0,35
Arrow-shape	NACA 6409 9%	12	1,33	0,88		4,48	2,96

According to the modelling results there will be chosen the wing with the following characteristics: the profile – NACA 6409 9%; the wingspan – 1 m; the mass of the wing – 0, 25 kg.

The slow growth of the lift force, which depends on the angle of attack, provides stability to the turbulence of the air flow. The range of angles of attack, ensuring a stable flight, is 6 - 12o. Large values can lead to the disruption of the airflow stream, the lift coefficient growth fall of the convertiplane [1, 3, 14].

4 The description and principle of movement of the convertiplane

In this part there will be analyzed the electro-mechanical system of the tricopter propeller drives (Fig.6) [9]. The propellers are driven by electric motors that generate the moments M_{EMi}, while the moment of external loads M_{Hi}, prevents the rotation of the propellers with angular velocities ω_i.

The restoring moment, acting on the frame from the electric drives of propellers, is determined by the value of the control voltages, coming from the automatic control system on the electric drives of the four propellers. And the level of voltages, generated by the power amplifier of the automatic control system, entirely depends on the type of controller.

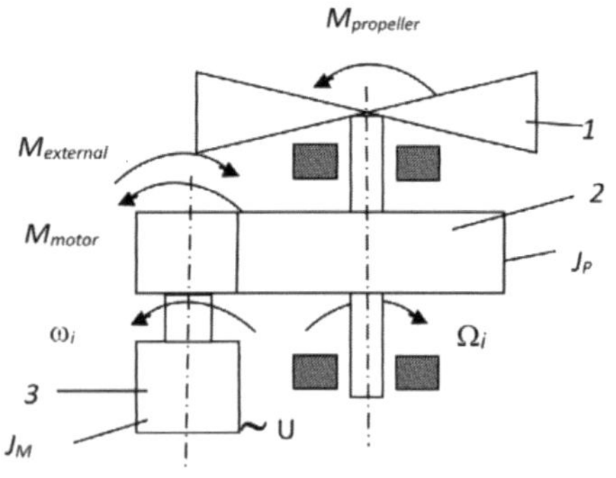

Fig. 6. The simplified scheme of the propeller electric drive: 1 – propeller; 2 – reducer; 3 – electric motor; U – control voltage; M_{motor} – the moment, produced by the electric motor; $M_{external}$ – the moment of the external loads of the propeller relative to the rotation axis; $M_{propeller}$ – the propeller torque; J_P, J_M – the moment of inertia of the rotor around the axis of the propeller and the axis of the motor, respectively; ω_i , Ω_i – the angular velocity of the motor and propeller, respectively.

The differential equations for the four propellers that describe interrelated electromagnetic and mechanical processes in the electromechanical system of the convertiplane propeller drives are the following:

$$\begin{cases} J_i \dfrac{d\omega_i}{dt} = M_{motor} - M_{external} \\ L_i \dfrac{di_i}{dt} + i_i R_i + c_E \omega_i = U_i \end{cases}, \quad (2)$$

where Li, Ri – the inductance and active resistance of the rotor winding; i – the current in the winding of the electromagnet; cE – the coefficient of proportionality, called the constant EMF of the motor; Ji – the reduced axial moment of inertia; Ui – the control voltage.

The moment produced by the motors:

$$M_{motor_i} = c_{M_i} \cdot i_i \quad (3)$$

where cMi – the constant of the motor torque.

The connection between the angular velocity of the motor ωi and the screw rotor propeller Ωi:

$$\omega_i / \Omega_i = N, \quad (4)$$

where N – the contact ratio of the reduction.

The moment of the external loads $M_{external}$ is defined as:

$$M_{external} = \frac{M_{propeller}}{N \cdot \eta} = \frac{\mu \Omega_i^2}{N \eta}, \quad (5)$$

where MBi – the moment of resistance to the propeller rotation, η – the efficiency of the electric drive that connects the mechanical energy of the engine axis and the propeller, μ – the load factor, reduced to the motor shaft.

Solving approximately the system of equations (2), taking into consideration (3), (5) we get equations, relating the angular velocity of propeller rotors and the control voltage supplied to the motors [8]:

$$\dot{\omega}_i = \frac{-\left(c_E c_M / R\right)\eta N^2 \omega_i - d\omega_i^2 + (c_M / R)\eta N U_i}{J_p + \eta N^2 J_M}, \quad (6)$$

where J_P, J_M – the moment of the rotor inertia around the axis of the propeller and the axis of the motor, respectively, d – the aerodynamic constant.

5 The synthesis of the energy efficient parameters of the electric drive

As the criterion of the energy efficiency there will be chosen the following value:

$$\varepsilon = |A_C| \frac{1}{\int_0^T W dt}, \quad (7)$$

where $\int_0^T W dt$ – the losses in electric drives, T – the time of motion, A_c – the work of the traction forces, the forces of gravity and resistance to movement at a given movement of the center of mass:

$$A_c = \int_0^T \bar{R} \cdot \dot{\mathbf{r}}_C dt, \quad (8)$$

where m – the total weight of the convertiplane, g – the gravitational acceleration, $\mathbf{r}_C = [X_C \ Y_C \ Z_C]^T$ – the vector, defining the position of the center of mass of the device, $\bar{R} = \bar{P} + \bar{Q} + \bar{G}$ – the main vector of external forces.

The value W that determines the losses in electric drives in the formula (7), will be presented like:

$$W = (\mathbf{u}^* - \mathbf{u})\mathbf{R}(\mathbf{u}^* - \mathbf{u}), \quad (9)$$

where **R** – the matrix of the given resistances of electric motors $\mathbf{u} = [M_1 \ M_2 \ M_3]^T$ – the vector of control actions, and $\mathbf{u}^* = [M_1^* \ M_1^* \ M_1^*]^T$ – the vector of control actions, which is necessary for obtaining the desired time dependences of the generalized coordinates $\mathbf{q}^*(t)$:

$$\mathbf{u}^* = \mathbf{B}^{-1}(\mathbf{H}\ddot{\mathbf{q}}^* + \mathbf{C} + \mathbf{G}), \quad (10)$$

$$\mathbf{u} = \mathbf{u}^* + \mathbf{B}^{-1}\mathbf{H}(\mathbf{K}_p\mathbf{e} + \mathbf{K}_d\dot{\mathbf{e}}), \quad (11)$$

where **H** – the generalized matrix of inertia of the mechanism, **C** – the vector of the generalized Coriolis and normal inertial forces, **G** – the vector of the generalized gravity, **B** – the matrix that connects the vector of control actions with the vector of the generalized forces, generated by the drives, $\mathbf{K}_p$ и $\mathbf{K}_d$ – the diagonal matrix of coefficients of the regulator:

$$\mathbf{K}_p = \begin{bmatrix} k_1 & 0 & 0 \\ 0 & k_2 & 0 \\ 0 & 0 & k_3 \end{bmatrix}, \mathbf{K}_d = \begin{bmatrix} k_4 & 0 & 0 \\ 0 & k_5 & 0 \\ 0 & 0 & k_6 \end{bmatrix}, \quad (12)$$

where k_j – the positive coefficients of the regulator ($j = \overline{1,6}$).

The values of the matrixes and vectors mentioned above, and the way of their finding are given in publications [12–15]. The general analysis methods of the equations of kinematics and dynamics of flying devices are shown in the following works [5–9]. The controller is described in the papers [7].

The currents in the windings of the electric drives are connected with the moments generated by the expression $C_\tau \eta I_i = M_i$, where I_i – the current in the windings of the *i*-drive, C_τ – the constant of the electric motors moments, η – the contact ratio of the reducer. The desired values of the currents I_i^* will be determined according to the formula: $I_i^* = M_i^*/(\eta C_\tau)$, and M_i^* will be received from the equation (10). The currents I_i^* determine the energy loss $W^* = \sum_{i=1}^{3} I_i^* R_i$, with the motion of an exoskeleton on the desired trajectory, where R_i – the resistance of windings of the *i*-drive. The value W^* corresponds to the minimum energy loss at a given trajectory of the gravity center of the convertiplane. The value W will be determined in a scalar form:

$$W = \sum_{i=1}^{3} (I_i^* - I_i)^2 R_i . \quad (13)$$

Taking into account (13) the matrix of given resistances **R** is:

$$\mathbf{R} = \begin{bmatrix} R_1/(\eta^2 C_\tau^2) & 0 & 0 \\ 0 & R_2/(\eta^2 C_\tau^2) & 0 \\ 0 & 0 & R_3/(\eta^2 C_\tau^2) \end{bmatrix}. \quad (14)$$

The formula (13) allows finding the additional losses in electric drives, resulting from the deviation of the real trajectory from the desired one. Minimizing Q it is possible to improve the energy efficiency of a convertiplane. The control system of a convertiplane is described in [7, 8] and consists of the following blocks: The "Generator of master controls" that defines the desired trajectory of the gravity center of the convertiplane, the "Block of solution" that evaluates $\mathbf{q}^*(t)$, implementing this trajectory, the comparator that calculates the error of control $\mathbf{e} = \mathbf{q}^* - \mathbf{q}$, the controller and the control object – the model of the convertiplane.

The method of the flying vehicle control should provide greater efficiency under the condition of minimizing the deviation of the center of mass of the convertiplane from the desired trajectory of movement.

6 The optimization of the control system parameters

As the vector of variable parameters there will be chosen $\mathbf{k} = [k_1\ k_2\ k_3\ k_4\ k_5\ k_6]^T$. The components of the vector k are defined on the intervals:

$$0 \le k_j \le k_{\max, j}, \quad (15)$$

where $k_{\max, j}$ – the maximum allowable value of the controller coefficient, $j = \overline{1,6}$.

The target function $J = J(\mathbf{k})$ will be defined as a weighted sum of two integral performance criteria:

$$J = \mu_1 J_1 + \mu_2 J_2 \quad (16)$$

$$J_1 = \int_0^T \left[(\mathbf{q}^* - \mathbf{q})^T (\mathbf{q}^* - \mathbf{q})\right] dt \Big/ \int_0^T \left[(\mathbf{q}^*)^T (\mathbf{q}^*)\right] dt, \quad (17)$$

$$J_1 = \int_0^T W dt \Big/ \int_0^T W^* dt, \quad (18)$$

where μ_1, μ_2 – the weight coefficients. By varying the coefficients, it is possible to obtain the regulators that can provide a greater precision, or greater energy efficiency. The criterion J_1 grows with the increase of the module of the control error **e**, and the criterion J_2 – with increasing of W value. Thus, minimizing J provides the decrease of both the errors of control and the losses.

The task of determining the optimal parameters of the controller consists of finding the vector $\mathbf{k}_m$, leading J to the minimum.

The methodology of determining the optimal parameters is focused on the method of sensing the space variable values which are determined by the values of criterion parameters at different points in space with a further solution of the problem of smooth surface's approximation of second order points.

7 Conclusion

As a result of the research, there has been proposed a model of a convertiplane, taking into account the optimal geometric parameters of the wings that meet the required aerodynamic characteristics and provide the increased maneuverability and the speed characteristics of this model. With the help of the XFLR5 software product there have been determined the coefficients of lift and resistance forces that will help determining the parameters of propellers and electric motors of the convertiplane.

There has also been formulated the task of the nonlinear optimization of the controller parameters of the convertiplane, implementing the flight in the vertical and horizontal planes. The objective function is defined as weighted sum of the two integral quality criteria related to accuracy and power consumption of the mechanical system. The optimization is accomplished by numerical methods of global optimization for different values of the weight coefficients of the objective function. It is shown that the optimal choice of the vector $\mathbf{k}_m$ reduces manifold the value of the objective function, increasing the accuracy of the automatic control system and reducing the power consumption.

References

1. G.S. Byushgens (Eds.). *Flight Dynamics* (Moscow, Mashinostroenie, 2011)
2. P.P. Afanasiev, I.S. Golubev et al. *Unmanned aerial vehicles* (Moscow, MAI, 2008)
3. N.F. Krasnov, Aerodynamics and wing and profile, In *The basic theory of aerodynamics* (Moscow, 1976)
4. M.L. Mil et al. *Helicopters* (Moscow, Mashinostroenie, 1967)
5. V.E Pavlovski, S.F. Jatsun, O.V. Emelyanova, S.P Stykanyova, *Unmanned vehicle with elements of artificial intelligence (UV-II-2015), proceedings of the second all-Russian scientific-practical seminar,* 99-106 (2015)
6. S.F. Jatsun, O.V. Emelyanov, A.I. Savin, S.P. Stykanyova, *News SWSU*, **14**, 55 (2015)
7. S.F. Jatsun, O.V. Yemelyanov, K.G. Kazaryan, *Unmanned vehicles with elements of artificial intelligence (UV-II-2016): proceedings of the third all-Russian scientific-practical seminar*, 147-157 (2016)
8. V.E. Pavlovsky, S. F. Jatsun, O.V. Emelyanova, A.V. Savitsky, *Robotics and technical cybernetics: scientific and technical journal*, 5, 49-57 (2014)
9. S.F. Jatsun, O.V. Emelyanova, A.I. Savin, *News of the SWSU. A series of engineering and technology*, **17**, 31-37 (2015)
10. The website "Podelise" http://tnu.podelise.ru/docs/index-385011.html?page=5
11. The website "Airfoil Tools" http://airfoiltools.com/
12. The website "Soviet helicopters" http://wiki.nashtransport.ru/wiki/Советские_вертолеты
13. G. Salazar-Cruz, Aerospace and Electronic Systems Conference (2008)
14. A. Oosedo, S. Abiko, S. Narasaki, A. Kuno et al. *2015 IEEE International Conference on Robotics and Automation (ICRA)*, 2326-2331 (2015)
15. S.F. Jatsun, S.V. Efimov et al. The patent for useful model No. 157967, IPC B64C 25/62. The air propulsion with variable thrust vectoring / (RF). No. 2015131037/11 (2015)

 DOI: 10.1051/matecconf/20179902003

Analysis of the Technological Parameters of the Heat Exchanger in the Heating Pipe

Vladimir Knyazev[1,*] and *Gregory* Bubnov[1, 2]

[1]Moscow Institute of Physics and Technology, Higher School of Systems Engineering, 141701 Dolgoprudny, Russia
[2]Moscow Technological Institute, 119334 Moscow, Russia

Abstract. The main purpose of this article is to analyze the selecting of technological parameters for the heat exchanger to improve the heat transfer and reduce the noise during operation in the heating pipe, which is used in the different systems of the planes and helicopters. In result of this study, the best technical parameters are found, considering different variations of deformation cutting heat exchanger pipes.

1 Introduction

Nowadays heating pipes are used in a wide range of aerospace systems, such as electronic cooling systems in aircraft, passive transmission cooling systems, gear box systems in helicopters and others [1]. Let's consider the principle of a heating pipe (thermosyfon) in a radiator based on heating pipe technology. The radiator consists of two independent circuits. Thermal energy comes from the trunk-heating pipe (processor) to the first circuit (heat exchanger pipe). Through the heat exchanger pipe, the energy is transmitted to the small amount of low-boiling liquid, located in the second circuit (sealed thermal panel). The pressure inside the thermal panel is reduced to 30-50 kPa. Therefore, low-boiling liquid begins to boil at 30-35 degrees Celsius. The resulting steam begins to rise up inside the thermal panel and condenses in the inner surface of the second circuit.

During condensation, thermal energy is transmitted to the surface of the second circuit and condensate flows downward due to gravity, after that the cycle repeats. According to that, thermal energy is transmitted from the trunk heating pipe to the surface of the panel, which respectively transmits this energy to the room. The difference between the temperatures of the heating transfer agent in the trunk heating pipe and the surface of the thermal panel is about 7-9 degrees Celsius. It depends on the various operating modes of the appliance.

The method of functional analysis and evaluation of significance functions [2] and the theory of multi-parameter optimization [3] are used in this article. The article [4] describes deformation cutting technology, which was used to design the microstructure surfaces on the heat exchanger pipe.

The analysis of the criteria for the heat exchanger pipe, and the choice of the optimal alternative to increase efficiency of the pipe during the operation will be the result of this work.

2 Deformation cutting technology for the heat exchanger pipe

Technology of the deformational cutting was chosen a perspective method to improve the heat transfer and reduce the noise in the heating pipe systems. The main purpose of the Deformational Cutting is to get the macro pattern in the form of pin fins, cells, meshes, v-grooves on such materials as Copper, Aluminum and their alloys, steel and plastics, with the opportunity to increase the surface area after the process up to 12 times. For the analysis four different alternatives of the Deformation cutting steel pipes were chosen (Fig. 1). The main characteristics of the pipes are shown in the Table 1.

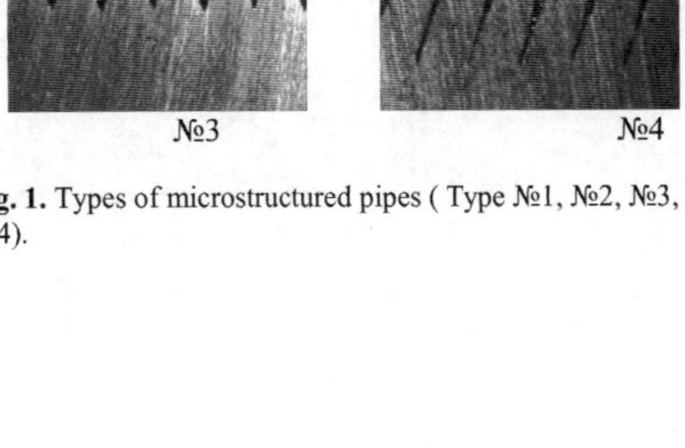

Fig. 1. Types of microstructured pipes (Type №1, №2, №3, №4).

* Corresponding author: vladimir.knzv@gmail.com

Table 1. Characteristics of the microstructured pipes

Parameters /Types	№1	№2	№3	№4
Step of the edges, mm	0,4	0,5	0,4	0,4
The average height of the structure , mm	0,95	0,75	0,65	1,15
The average width of the intercostal gap, μm	130	250/1 00	100	40
Deviation from the vertical ribs	0	0	15	15

3 Functional analysis

Functional analysis (Table 2) will help to define the most significant parameters, which could improve the heat transfer and reduce the noise in the heating pipe systems

Table 2. Functional analysis

Requirements (Upper level)	Requirement s (Derivative)	Functions
Ensure the film boiling	Reduce the noise	F1 Ensure the optimum width of the intercostal gap
Increase the heat transfer	Increase heat transfer of the appliance	F2 Ensure the optimum height of the structure
		F3 Ensure the optimum step the edges
Ensure the minimal changes in diameter of the pipe	Increase the build capacity	F4 Ensure the optimum deviation from the vertical ribs

Use the matrix N2 (Table 3) to reduce the number of functions. The N2 matrix has the dimensions correspond to the number of functions. At each intersection of rows and columns the expert assessment of the pair interaction is placed. If the functions are equivalent, the sign "="is put. If the function value is stored in the line is greater than the function value stored in the column, then the sign ">" is put. Otherwise, the sign "<" is put.

Table 3. N2 matrix

	F1	F2	F3	F4
F1	=	>	=	>
F2	<	=	>	>
F3	=	<	=	>
F4	<	<	<	=

Replace the symbols < , = , > with the numbers 1, 2, 3, and sum the numeric values in each row (Table 4).

Table 4. N2 matrix (numeric values)

	F1	F2	F3	F4	Total
F1	2	3	2	3	10
F2	1	2	3	3	9
F3	2	1	2	3	8
F4	1	1	1	2	5

Each line must be multiplied by the column vector sum in order to calculate the absolute priority function for this criterion.

Calculating the priority ranking of each function provides an estimation of the importance of this function (Table 5). By itself, this number does not make any sense and depends only on the scale of adopted assessments. Relative rating values indicate the significance of the function

Table 5. Significance of the function

№	Absolute value of the function	Significance of the function %
F1	100	37,0%
F2	81	30,0%
F3	64	23,7%
F4	25	9,3%
Total	270	100%

Implement the following notation from 0 to 1 (from worst to the best value of the parameter) (Table 6).

Table 6. Relative values of the optimization parameters

Parameters/ Alternatives		№1	№2	№3	№4	Weight
1	The average width of the intercostal gap	0,5	0,7	1	0,2	0,4
2	The average height of the structure	0,7	0,6	1	0,9	0,3
3	Step of the edges	0,7	1	1	1	0,2
4	Deviation from the vertical ribs	0,5	0,7	0,7	1	0,1

Germeier's method was chosen as the most suitable, among the large variety of multi-parameter optimization methods [3]. Optimization criteria were defined by the requirements to the outer surface of the heat exchanger pipe: the average width of the intercostal gap, the average height of the structure, step of the edges, and deviation from the vertical ribs. It is not feasible to sum the values directly to optimize the parameters of each alternative, because they have different dimensions. For that, these value parameters must be expressed in the same units. It is convenient to express them in a fraction of the best. In order to do this, the following formula is used:

$$X = \frac{X_{best} - X_i}{X_{best} K_i}.$$

Where X — is the optimization value, expressed as a fraction of the better, X_i – is the value of the parameter optimization in real terms, X_{best} – is the value of the parameter optimization, the best among the considered alternatives, K_i — is the value of the weighting factor for the i-th optimization parameter. The experts, depending on the particular importance of the optimization parameter, determine K_i values.

How close the value of the integral Germeier criterion to zero, determines the optimal alternative (Table 7).

Table 7. Germeier's criteria

	Parameters/ Alternatives	**№1**	**№2**	**№3**	**№4**	**Weight**
1	The average width of the intercostal gap	1,3	0,8	0,0	2,3	0,4
2	The average height of the structure	1,0	1,3	0,0	0,3	0,3
3	Step of the edges	1,5	0,0	0,0	0,0	0,2
4	Deviation from the vertical ribs	5,0	2,0	2,0	0,0	0,1
Germeier's criteria		8,8	4,1	2,0	2,6	1

4 Conclusion

First, deformation cutting technology for the heat exchanger pipes was described, which improves the efficiency of the heat exchange and reduces the noise during operation in the heating pipe systems. After that, multi-criteria analysis was held, to select the alternatives of the deformation cutting steel pipes through the functional analysis and multi-parameter optimization of Germeier's method. Based on this analysis, the best technical solution to increase the heat transfer efficiency and reduce the noise during operation in the heating pipe systems is the alternative №3 (Fig. 1).

References

1. D. Peter, D. Reay, *Heat pipes* (Elsevier, 2012)
2. B. Lightsey, *Systems engineering fundamentals* (Defense Acquisition Univ, 2001)
3. Y.B. Germeier, *Introduction to the theory of operations research* (Moscow, Nauka, 1971)
4. N.N. Zubkov, *Russian Engineering Research*, **35**, 859-863 (2015)

Shape of the Heat Exchanger Pipe in the Heating and Cooling Systems Based on Heating Pipe Technology in Aircraft

Vladimir Knyazev[1,*], and *Evgeny* Nikulchev[2, 3]

[1]Moscow Institute of Physics and Technology, Higher School of Systems Engineering, 141701 Dolgoprudny, Russia
[2]Moscow Technological Institute, 119334 Moscow, Russia
[3]Moscow Technological Institute MIREA, 119454 Moscow, Russia

Abstract. The main purpose of this article is experimental research aimed to find the required shape of the heat exchanger pipe manufactured by deformation cutting technology for increased efficiency during operation in the heating and cooling systems based on heating pipe technology in aircraft. In the result, recommendations were made to use the different shapes of the heat exchanger pipe in the heating and cooling systems based on heating pipe technology in aircraft.

1 Introduction

There are many ways to use heating pipe technology in the aircraft's equipment. The main principle of heating pipe technology is described in the book [1]. The article [2] describes the in-flight entertainment cooling system, which is used on commercial boards and provides audio, video, internet, multimedia, phone services. This system is based on the heating pipes and could be installed under each passenger's seat in a modern aircraft. The patent [3] describes the passive cooling system for the aircraft's transmission using the mast mounted heating pipes. The patent [4] shows a cooling system based on heating pipes for helicopter gearbox. The heating pipe is also used in heating systems, such as passive anti-icing systems in aircrafts.

One of the methods to increase heat transfer in the heat exchanger pipe and simultaneously save the required dimensions of the pipe is deformation cutting technology. The article [5] describes requirements and technology for fin application to the long pipe on the basis of hybrid deformation and cutting. In another article, [6] the results are described from an experimental research of heat transfer in the microstructured surfaces obtained by using the deforming cutting method.

The aim of the experiment is to evaluate the effectiveness of deformation cutting technology used in manufacturing of the heat exchanger pipes in the heating and cooling systems based on a heating pipe technology in aircraft.

2 Types of the heat exchanger pipes

Technology of the deformational cutting was chosen as a perspective method to improve effectiveness in the heat exchanger pipes. All samples have following parameters: 20 mm diameter, 2.5 mm wall thickness, made of steel B20. The length of all the samples is 1100 mm. The inner pipe surface area is 0.0471 m2. Four samples were provided for the experiment; with different outer surfaces are obtained using deformation cutting technology (Fig. 1). The main characteristics of the pipes are showen in the Table 1.

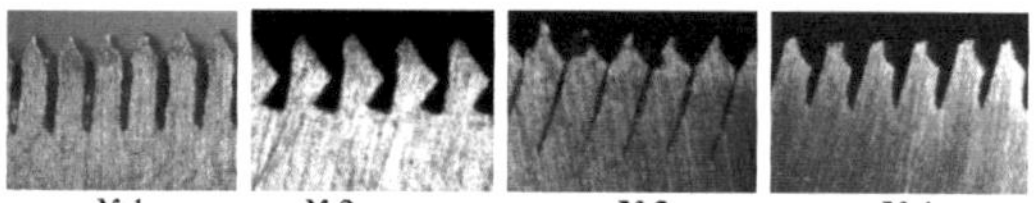

Fig. 1. Types of microstructured pipes (Type №1, №2, №3, №4).

Table 1. Characteristics of the microstructured pipes

Parameters /Types	№1	№2	№3	№4
Step of the edges, mm	0,4	0,5	0,4	0,4
The average height of the structure , mm	0,95	0,75	0,65	1,15
The average width of the intercostal gap, μm	130	250/100	100	40
Deviation from the vertical ribs	0	0	15	15

3 Experimental setup and measurement procedure

The experimental setup is imitating the operation of the two phase heating pipe circuit. It consists of two independent circuits. Thermal energy comes from the trunk-heating pipe to the first circuit (the heat exchanger pipe). Through the heat exchanger pipe, energy is

* Corresponding author: vladimir.knzv@gmail.com

 DOI: 10.1051/matecconf/20179902004

transmitted to the small amount of low-boiling liquid, located in the second circuit. Pressure inside the thermal panel is reduced to 30-50 kPa. Therefore low-boiling liquid begins to boil at 30-35 degrees Celsius. The resulting steam begins to rise up inside the thermal panel and condenses in the inner surface of the second circuit. During condensation, thermal energy is transmitted to the surface of the second circuit and condensate flows downward due to gravity, after that the cycle repeats. According to that, thermal energy is transmitted from the trunk heating pipe to the surface of the second circuit, which respectively transmits this energy.

The experiments were carried out at the setup (Fig. 2) consisting of the following components. In the first circuit, hot water flows from the local heating system into the testing pipe (3). The thermal sensors (9) measure the input and output temperatures in the testing pipe. The flow meter (10) measures the water flow in the testing pipe. In the second circuit there are: condenser (1), where vaporized low-boiling liquid condenses, transparent pipe (2) made of plastic, pipelines (4), valves (5) for flow control, condensate container (6), vacuum gaskets (7), supply vessel (8) to support the required level of liquid in the transparent pipe during the experiment, pressure sensor (11) and vacuum pump (12) for setting required pressure in the second circuit.

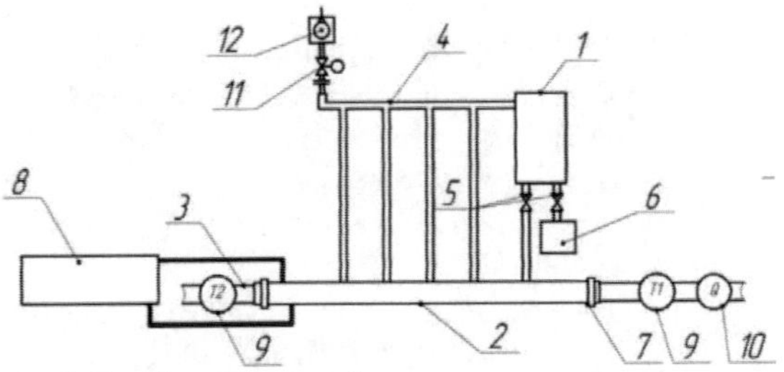

Fig. 2. Experimental setup scheme.

In the testing pipe, the stationary flow rate and temperature of water are set. The second circuit is evacuated from vapor pressure of low-boiling liquid, which is filled into the transparent pipe. In this case, it was isopropyl alcohol. The condensate container's valve is closed; the system has reached the stationary operating condition.

Next the condensate container's valve is opened and the transparent pipe's valve is closed at the same time. Condensate begins to drain into the condensate container. Measured values during the experiment are: the mass of evaporated isopropyl alcohol depending on the time, input and output temperatures of the water in the testing pipe, the water flow rate, pressure and level of isopropyl alcohol in the transparent pipe.

The graphs of the mass of evaporated isopropyl alcohol depending on time are obtained based on the results of the experiment. The graph is divided into the sections, corresponding to the different levels of isopropyl alcohol in the transparent pipe. After that, the angular inclination dm/dt is derived at each site. Thermal power of the pipe is calculated according to the formula:

$$W = q\frac{dm}{dt}.$$

The pipes with different outer surfaces are compared to the standard pipe with the smooth surface. The relation between the thermal power and temperature in the testing pipe are obtained for different immersion levels of isopropyl alcohol in the testing pipe. The most revealing immersion level (Figure 3) was found after several iterations and it was studied more thoroughly.

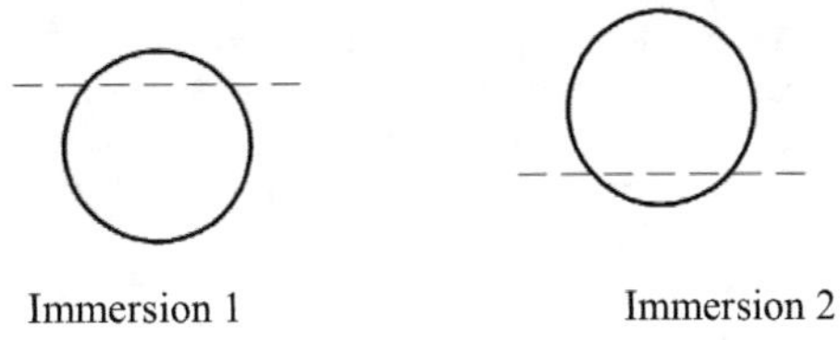

Fig. 3. Immersion levels.

4 Results

For the immersion level 1, all of the test samples worked loudly in the nucleate boiling regime. The pipes manufactured by deformation cutting technology showed the increased heat output, more than 15-20% higher than the standard pipe. The experimental data is shown in the Fig. 4.

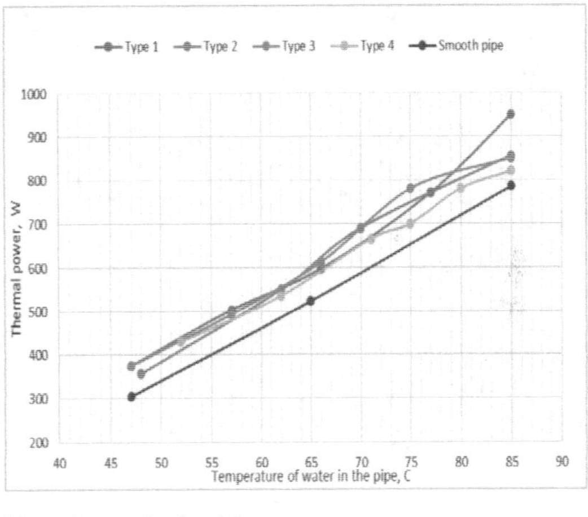

Fig. 4. Immersion level 1.

For the immersion level 2, all the pipes manufactured by deformation cutting technology showed significantly greater heat output (from 30% to 100%, depending on the temperature) than the standard pipe with the smooth surface. This is because the pipes manufactured by deformation cutting technology have high thermal conductive capillary channels. Evaporation takes place within the entire surface of the pipes. At the same time, different types of surfaces display different features.

For the type 1, at the temperatures up to 65 degrees Celsius, the pipe actively evaporates alcohol from the entire surface. Film boiling occurs, which significantly reduces the noise during operation. In the range of 65-75 degrees Celsius the capillary channels begin to dry out, parts of the pipe are no longer wet. In result, heating power does not increase with increasing temperatures. Film boiling continues at the temperatures of over 75 degrees Celsius, the process follows to the active phase of the nucleate boiling. In this case, since the gap between the pipe and the inner surface of the transparent pipe is sufficiently small, and the formed spray covered the entire surface of the pipe, the capacity significantly increases. For the types 2 and 3, at the temperatures up to 65 degrees Celsius, the pipes have similar specifications as the type 1. At the temperature of over 65 degrees Celsius, the nucleate boiling begins, so the power failure is not observed. For the type 4, the surface evaporation continues to up to 80 degrees Celsius, and the capillary channels do not dry out. At the high temperatures (more than 80 degrees Celsius), the process follows to the nucleate boiling phase. The experimental data is shown in the Fig. 5.

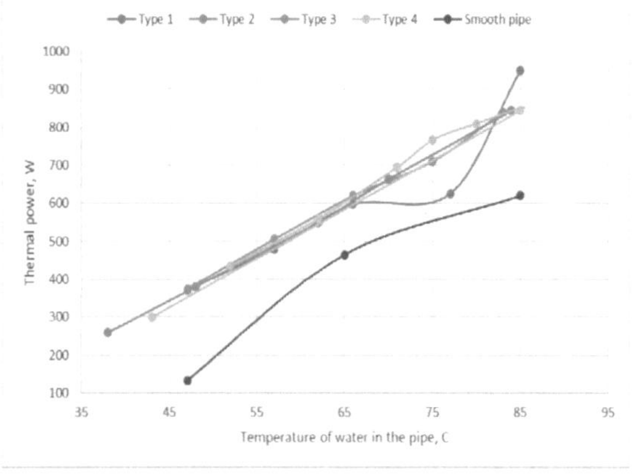

Fig. 5. Immersion level 2.

5 Conclusions

This article described the experiment studying features of the pipes manufactured by deformation cutting technology for the heat exchanger pipes used in the heating and cooling systems, based on heating pipe technology in aircraft. For the immersion level 1, the pipes manufactured by deformation cutting technology showed the increased heat output, more than 15-20% higher than the standard pipe. For the immersion level 2, the sample 1 has the power failure observed in the heating water temperature range of about 65-75 degrees Celsius. The samples 2 and 3 can significantly improve the performance of the heat exchangers and reduce the noise over a wide range of temperatures. Sample 3 showed the best heat transfer features, among all samples.

The heat exchange surface, similar to the surface 3, can be recommended for using as a part of a system based on a heating pipe technology. The heat exchange surface obtained by deformation cutting technology may be used in the heating and cooling systems, based on a heating pipe technology in aircraft to increase the heat transfer and reduce the noise during operation.

References

1. D. Peter, D. Reay, *Heat pipes* (Elsevier, 2012)
2. Sarno C. et al., *Applied Thermal Engineering,* **51**(1), 764-769 (2013)
3. R.T. Ehinger, M.A. McGlaun, *Passive cooling of transmission using mast mounted heat pipes*, Patent. 8985517. (2015)
4. M.A. McGlaun, K.M. Jackson, *Helicopter gearbox auxiliary cooling system*, Patent 9272777 USA. (2016)
5. N.N. Zubkov, *Russian Engineering Research*, **35**(11), 859-863 (2015)
6. I.A. Popov, *Thermal Engineering*, **60**(3), 157-165 (2013)

DOI: 10.1051/matecconf/20179902005

Application of Queuing Systems with Ultimate Reliability to Optimize the Operation of Drones

Anna Kovalenko[1] and *Gennagi* Rogachev[1, *]

[1]Samara state technical university, 443100, Samara, Russia

Abstract. Multi-criteria problem of optimization of maintenance rate for drones is set and solved. Analytical characteristics of a single-server queueing system with ultimate reliability and imitation models are applied. The analysis of economical and reliability characteristics dependence on some parameters is made.

1 Introduction

Drone is a complex object, result of solving the various technical and economic problems [1]. For sure, the primary problem is to develop the system of automatic drone control. In particular, the synthesis of control algorithms to stabilize flight trajectories is important [2]. However, like all complex technical devices with a large number of mechanical parts being a subject to constant deterioration, drones undergo break downs. One should also note the high cost, the responsibility for the quality of the problems solution and the effects of the environment [3]. Due to these reasons, the consideration of drone reliability in terms of mechanical failures and the solution of the drone survivability problem are quite relevant.

One of the methods of increasing the drones' reliability is to provide control at the design stage in the process of creating the devices [4] and to plan scheduled maintenance. In this paper, the optimization of maintenance rate of a drone operating according to the model of a single-server queueing system [5, 6] is made.

The main types of drone hardware failures are 1) Batteries discharge; 2) Engine failure; 3) Mechanical damage (improper motor shaft rotation, background noise and vibration, shear beam, etc.)

Preventive measures for the drone hardware include visual inspection to detect failures, maintenance (charging batteries, engine replacement or repair, shift-rays, etc). The essence of the strategy of drones maintenance described in [5] is the periodic (counting operating time) maintenance carried out after the end of task during which certain value of device operating time is reached.

2 Problem setting

Consider the operation of Geoksan 201, a drone of aircraft type. These drones are best suited for use over large areas of extended objects and shooting at a considerable distance. Geoskan 201 is capable of flying up to 180 minutes with a payload (camera), shooting up to 22 sq.km per one flight. However, due to the design features, it has to move constantly. It will not be able to hang in a necessary point or work in a confined space. Therefore, the device is effectively applied to shooting agricultural areas.

Suppose that the drone tasks arrive in a stream with the average inter-arrival time of 7 hours. The duration of task completion has a third-order Erlang distribution with the average of 5 hours (meaning that one task can include multiple drone flights). During a flight, a drone failure can occur. Then immediately the detection of the device, its transportation, failure analysis and restoration are carried out. The device failure is assumed to occur when the operating time reaches the random value having the Erlang fourth-order distribution with an average value of 300 hours. The operation time countdown starts at the moment of the first task income after restoration. The duration of restoration is a random variable with the average of 4 hours. Maintenance begins after the completion of the task during which device operation time has a predetermined level τ. Both after maintenance and restoration reliability characteristics are completely renewed. The duration of maintenance is a random variable with the average of 1 hour. All the random variables are considered to be independent. The income per completed task is $C_{ser} = 2000$ money units, the cost per time unit of restoration and maintenance are $c_a = 1200$ money units/hour and $c_p = 200$ money units/hour respectively.

3 Optimization of the drone maintenance rate

The following criteria, analytically defined in [5, 6], are taken into consideration:

* Corresponding author: grogachev@mail.ru

1) a positive criterion - average specific income $S(\tau)$ per unit of calendar time;
2) a negative criterion - average unit cost $C(\tau)$ per time unit of correct device operation;
3) a positive criterion - final probability $p^0(\tau)$ of a device standby mode;
4) a negative criterion - final probability $p^-(\tau)$ of maintenance or restoration; $p^-(\tau)$ also should not exceed a value p^-;

$$\begin{cases} S(\tau) \to \max\limits_{\tau\in T}, \\ C(\tau) \to \min\limits_{\tau\in T}, \\ p^0(\tau) \to \max\limits_{\tau\in T}, \\ p^-(\tau) \to \min\limits_{\tau\in T}, \\ T = \{\tau \mid p^-(\tau) \le p^-\}. \end{cases}$$

If $p^- = 0{,}05$, the restriction of the problem is $\tau \in T,\ T = \{\tau \mid p^-(\tau) \le 0{,}02\}$. In our case, it means $\tau \in (0;\infty)$.

One can show that in this problem the set of Pareto-optimal points can be obtained by maximizing the additive or multiplicative convolutions of partial criteria on the set T under different criteria weight coefficients.

Consider here an example of solving the optimization problem under equal criteria weight coefficients.

4 Results of the optimization problem solution in Maple and Anylogic

Without maintenance economical and reliability operation indexes are $S(\infty) = 42{,}231$ money units/hour, $C(\infty) = 9{,}639$ money units/hour, $p^0(\infty) = 0{,}19522$, and $p^-(\infty) = 0{,}00797$.

Under maintenance according to the strategy described in the problem setting, optimal values of specific criteria are $S(\tau^S_{opt}) = 51{,}036$ money units/hour under $\tau^S_{opt} = 81{,}395$, $C(\tau^C_{opt}) = 2{,}402$ money units/hour under $\tau^C_{opt} = 81{,}066$, $p^0(\tau^{p0}_{opt}) = p^0(\infty) = 0{,}19522$ under $\tau \to \infty$ and $p^-(\tau^{p-}_{opt}) = 0{,}00594$ under $\tau^{p-}_{opt} = 206{,}306$.

Pareto set consists of the sets of specific criteria values corresponding the values of τ from semi-infinite interval $\Pi = \tau \in [\tau^C_{opt};\infty) \bigcup T = [\tau^C_{opt};\infty)$.

An additive convolution of specific criteria under equal weight coefficients $v_s = v_c = v_p = 1/4$ reaches its maximum under $\tau^V_{opt} = 170{,}273$ hours; and $S(\tau^V_{opt}) = 49{,}757$ money units/hour, $C(\tau^V_{opt}) = 3{,}453$ money units/hour, $p^0(\tau^V_{opt}) = 0{,}19005$, $p^-(\tau^V_{opt}) = 0{,}00604$. At this point the values of specific criteria in absolute terms differ from the optima of specific criteria for 2,5% ($S(\tau)$), for 43,7% ($C(\tau)$), for 2,6% ($p^0(\tau)$) and for 1,7% ($p^-(\tau)$).

The multiplicative convolution of specific criteria under equal weight coefficients $W_s = W_c = W_p = 0{,}25$ reaches its minimum under $\tau^W_{opt} = 146{,}221$ hours; and $S(\tau^W_{opt}) = 50{,}262$ money units/hour, $C(\tau^W_{opt}) = 3{,}040$ money units/hour, $p^0(\tau^W_{opt}) = 0{,}18958$, $p^-(\tau^V_{opt}) = 0{,}00604$. At this point the values of specific criteria in absolute terms differ from the optima of specific criteria for 1,52% ($S(\tau)$), for 26,52% ($C(\tau)$), for 2,82% ($p^0(\tau)$) and for 5,75% ($p^-(\tau)$).

5 The dependence of the optimal maintenance rate on some parameters of device

Fig. 1, 2 represent graphs of economical characteristics against τ under different values of c_p.

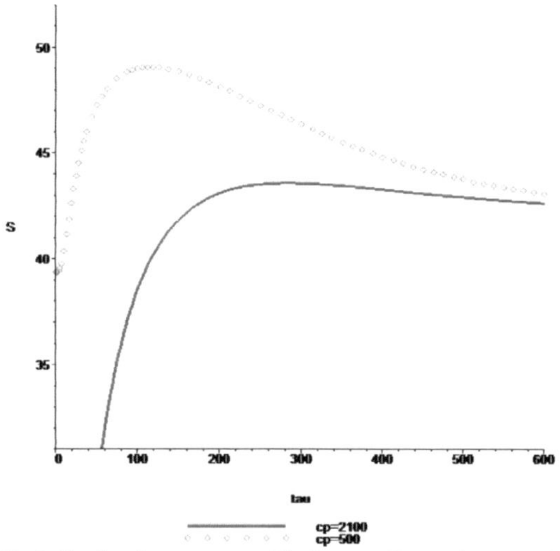

Fig.1 Graph of average specific income $S(\tau)$ against τ under different values of c_p.

With the increase of the cost c_p per maintenance time unit:
- the optimal maintenance rate increases; it is more evident for specific economical criteria and less evident for multi-criteria optimization;
- the value of the specific average income per time unit decreases;
- the value of the specific average cost per time unit of correct device operation increases;
- the optimum of specific criteria becomes less evident (for the value of approximately $c_p = 3000$ the cost function becomes monotonous, for $c_p = 3650$ the income function does).

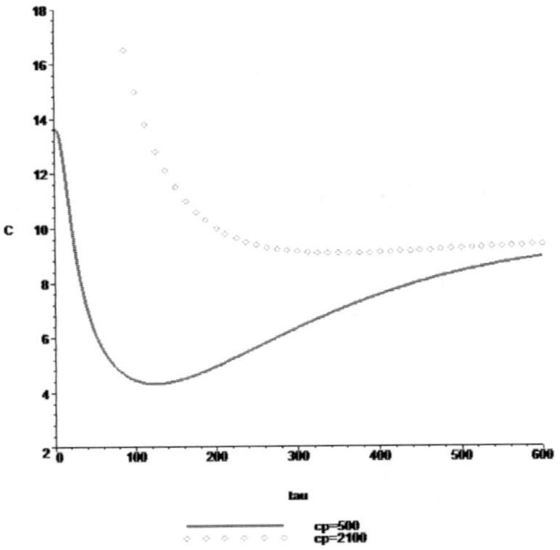

Fig.2 Graphs of average specific income $S(\tau)$ and average specific cost $C(\tau)$ against τ under different values of c_p.

With the increase of the intensity of income stream λ:
- the optimal maintenance rate does not change significantly;
- the value of the specific average income per time unit increases;
- the value of the specific average cost per time unit of correct device operation increases;
- the optima of specific economical criteria become more evident, while the optimum of final probability of device maintenance or restoration becomes less evident (for $\lambda = 0{,}33$ this criterion reaches its minimum under $\tau \to \infty$).

6 Simulation in Anylogic

Discrete event simulation of the device operation is carried out with Anylogic software. The operation of this model is shown in Fig. 3. In the upper left corner there are model parameters. To the right of the box model, you can see the current results of the calculations. With this model it is possible to analyze the stationary, non-stationary characteristics of the system, to carry out optimization experiments and make statistical analysis and visualization (as in Fig. 4).

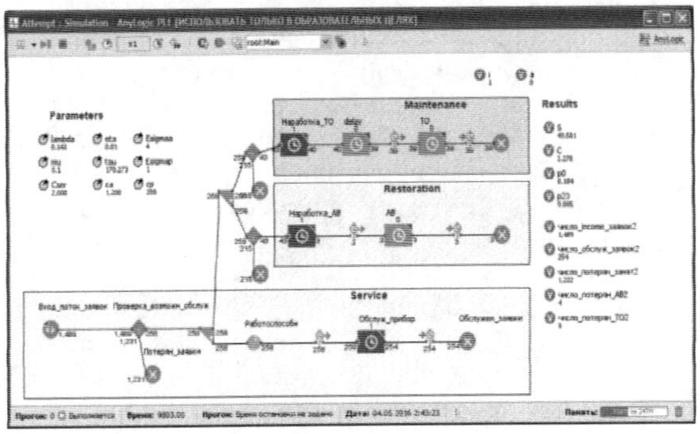

Fig. 3. Simulation process of drone operation.

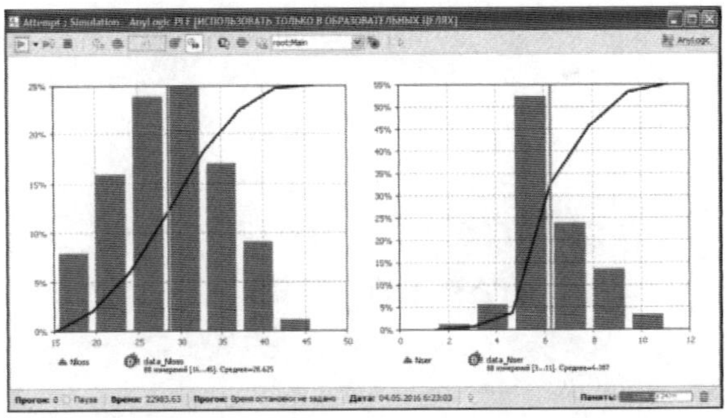

Fig. 4. Histograms of the numbers of lost and completed tasks (correspondingly) per regeneration period.

The optimization experiment in Anylogic can give us the values of optimal device rate for most of the criteria. However, you need 50 iterations with 20 replications (for the specific average income criterion) and much more for the other criteria and multiplicative convolutions to get results, coinciding with those calculated according to the analytical formulas up to 0,03.

7 Conclusions

The high cost of drones, their accessories, and restoration, as well as the impossibility of direct inspection while their operation, results in the need for maintenance to improve hardware reliability. Carrying out drone maintenance according to the strategy proposed increases the average specific income per time unit for at least 15% and reduces the average specific cost per time unit of correct device operation more than 1,5 times. The final probability of device standby mode is reduced not more than for 2,7%, and the final probability of maintenance or restoration slightly decreases.

This work was financially supported by Ministry of Education and Science of Russian Federation (basic part of government contract).

References

1. A. N. Katulev, *Operation Research: Principles of Decision Making and Safety Ensurance* (Tver', 1999)
2. A. A. Antanevich, Yu. F. Ikuas, L. A. Lobaty, Science and Technics, **5** (2010)
3. A. V. Betskov, I. V. Prokop'ev, *Reliability and Quality of Complex Systems*, **2**(6) (2014)
4. I. I. Ogoltsov, N. B. Rozhnin, V. V. Sheval', *Izvestiya TulSU. Technical Sciences*, **1** (2012)
5. A.I. Peschansky, A.I. Kovalenko, *Automatic Control and Computer Sciences*, **50(6)** (2016)
6. A. Kovalenko, G. Rogachev, *ITM Web of Conferences*, **6** (2016)

Problems of Formation of Diagnostic Features in the Diagnosis of Aircraft Engines

V. I. Pronyakin[1,*]

[1]Bauman Moscow State Technical University, 105005 Moscow, Russia

Abstract. The article is devoted to the evaluation of current technical condition of aircraft engines. Deals with the choice of the detection method of diagnostic features required for degradation assessment, emergency protection and detection of incipient defects on the example of cyclic machines and mechanisms. For the formation of diagnostic features in the diagnosis of aircraft engines use different physical effects (vibration, shock, heat radiation, electrodynamic and thermal processes, wear debris in oil, etc.). Classification of defects and requirements for the development of diagnostics systems is formed based on them. The article describes the requirements for diagnostic signs. The article provides a promising phase method that allows obtaining stable diagnostic characters in exploitation. The result of applying the method is shown. Diagnostic signs are formed. In mathematical modeling it was used the traditional theory of the description of rotary mechanisms. The data obtained are compared with experimental data.

1 Introduction

Currently, the task of assessing the technical condition of mechanical and electromechanical systems and predict their trouble-free operation is solved, as a rule, on the basis of a large volume of statistical data and failure analysis.

For complex technical objects (gas turbine engines, turbines, etc.) are dominated by expert assessment, which are determined by technical level experts.

This article focuses on the consistent diagnostic features necessary for the registration of the degradation and identification of incipient defects in functional diagnostics.

To assess the technical condition used the regular parameters and physical effects that occur in the process of working of gas turbine engine (GTE) (vibration, shock, heat radiation, electrodynamic and thermal processes, the products of wear in the oil, etc.).

It should be noted that these physical phenomena for high-speed engines are slow.

The processes associated with rotary engine, proceed quickly. The evaluation, built on the operating elements of the device failure, do not provide reliable emergency protection. The refusal of the engine during flight is unacceptable. For the exception of emergency needs to be performed on the ground assessment of the current technical condition and to provide reliable emergency protection in flight. While the kinematic parameters of movement of the machine elements resulting from the worker processes, usually for diagnostic purposes are not controlled. Possibilities of widely used parametric diagnostics grow due to the increase in the number of controlled parameters and increase the potential of computing machinery. Regular parametric diagnostics registers are already visible deviations from the standard values in the machine. This demonstrates the need of prevention or repair.

Currently, the diagnoses of GTE are widely used statistical methods, analysis of failures and their consequences for risk assessment. Life classification of defects and development of requirements for a diagnostic system are running on this database. Also used maintenance based on reliability theory. However, the statistical definition of probability of failure does not guarantee complete safety (for example, GTE in aviation). It is necessary to carry out continuous monitoring of current technical condition and reliable emergency protection complex and expensive technical objects.

The analysis of the traditional methods of obtaining information on the operation of the rotary mechanisms, including the vibration analysis showed that the main reasons for the lack of effective information-metrological maintenance of operation of machines and mechanisms include the following factors:

1. Methods of amplitude (e.g., vibration monitoring) to obtain information to solve the problem of assessing the current technical condition, the problem of the guarantee of emergency protection. Methods do not allow changing from the system of preventive maintenance systems the maintenance to systems on the current technical condition.

2. The metrological level of the measuring instruments used for diagnostics of machines and mechanisms, currently, medium, or low. In practice of metrological assurance, relative accuracy is a few percent. In this regard, the necessary informative content

* Corresponding author: vip-u@yandex.ru

of measurement information is not provided. Monitoring characteristics of vibro-acoustic processes is the main approach in the diagnosis of operating machines and mechanisms. Assessment of the vibration state of large machinery, identify the sources of vibration and its maximum remains fundamentally important, but unresolved to the full task. Low metrological level of the results of measurement of vibration parameters (the relative error from 1 to 10 %) allows to obtain informative quantitative individual characteristics required to assess the current technical condition, determining incipient defects and prognosis. Therefore, it remains a problem obtaining a stable diagnostic characters (or slowly changing) to assess the performance of rotor systems at all stages of the life cycle. Vibration measurement for functioning mechanisms with distributed mass allows obtaining statistical characteristics. Automatic detection of individual defects implemented. Applied monitoring systems measure the current values of parameters and provide automatic responses when thresholds are reached.

Expert evaluations prevail at the present time. They are subjective in nature. Assessments do not provide the effective operational solutions to perform the control actions of emergency protection.

3. Effective diagnosis of low speed and low-speed mechanisms with low vibration, high-speed systems and devices with distributed mass, etc. is not provided. The information collection system, registered vibroacoustic processes are improved. Methods of information processing are improving mostly.

4. Determining the relationship between measurements and peculiarities of functioning of the structural elements is a fundamental problem of traditional methods. Reliable correlation between measurement results and design, to obtain a quantitative estimation of the degradation components of the device, using the applied physical quantity (e.g., velocity and vibration acceleration), is unable to install. In this regard, the creation of automated and automatic systems of diagnostics and emergency protection of the functioning of machines and mechanisms remains an important issue.

Stage classification of the defects, currently implemented in the analysis of failures and technogenic accidents in the process of developing methods and tools for assessment of technical state of machines and mechanisms, is required. The detection of defects and determination of diagnostic features is made on the basis of the obtained results. Mathematical modeling of mechanisms using computational experiments for the simulation of defects are used to a lesser extent in the practice of diagnosis.

5. In the analysis of rotary mechanisms are widely used spectra of oscillatory processes. Spectra of oscillatory processes provide basic diagnostic information, but have a weak immunity. The main problem the most common amplitude methods is that the acoustic signal contains all the information about the interaction of parts of the working mechanism, including its natural vibration background. But how to retrieve information about a particular node, how to localize failures and to obtain quantitative characteristics of the changes - that's the question. The complexity of the defects is determined by high level of noise and a small change of the useful signal. The structure of the spectra changes over time (particularly for long intervals of use) in connection with a radical change of vibration parameters and their use for assessment of the evolution of the device, its technical condition, obtain trends, and forecast is very problematic. As a rule, a significant increase in the overall level of the signal or its individual components says only gross changes in the condition diagnosed mechanism leading to the loss of its functionality. The localization of developing problems requires the determination of the relationship of vibro-acoustic signal with the kinematics and dynamics of structural elements of the cyclic mechanism. There are positive results for concentrated systems (for example, a bearing). The main problem is distributed mechanical and Electromechanical systems (e.g., turbo - and hydroelectric generators, gas turbines, etc.).

Thus, in connection with the with-ranks of failed traditional methods and means to fully solve the problem of obtaining information about features of rotary mechanisms and to solve the problem of obtaining stable diagnostic features.

2 The requirements for diagnostic signs

Obtaining information about the functioning of the device is the main task of technical diagnostics. Processing of results of measurement of various physical quantities, selected in connection with the task, is a source of information. Thus it is necessary to solve a number of problems related to the task and metrological support and to determine:

- measurement accuracy associated with the level of required technical information,
- primary transducers with operating conditions, and testing;
- number of measuring channels
- the relationship of the measurement data with the design and controlled process product
- algorithms of processing of measuring information

etc.

The selected method of diagnosis determines the controlled physical quantity, measuring instruments and methods of mathematical processing of measurement results to correlate them with the design of controlled nodes and identify diagnostic features. Currently, the analysis of failures and identification of their Association with defects of the device are performed. It's a long process, which is unacceptable to ensure trouble-free operation of complex and expensive technical objects. Analysis of the impact of failures, breakdowns, industrial accidents and catastrophes runs of the decade, but to solve the problem of estimating the current technical status effectively fails.

2.1. The traditional approach

It is accepted that in systems of technical diagnostics use the principle of deviations (principle of Salisbury), and therefore it is proposed to use the characteristics of a "reference product or a mathematical model of the standard." However, the inevitable spread of parameters due to manufacturing errors, Assembly, adjustment and the impact of external influences. When high measurement accuracy is necessary to obtain detailed information, register the variation of parameters and therefore, the "reference values" may not have nominal value. In this connection there are ranges and algorithms that contain a number of consistent characteristics for the evaluation of the mechanism work. Currently various diagnostic methods are being intensively developed (neural networks, statistical analysis, information theory, theory of pattern recognition, the Bayesian method, etc.). But they do not provide a solution to the problem of assessing the current technical condition of operating machines and mechanisms. Statistical diagnostic methods and the Bayesian method, implemented on the basis of statistical information. They have different rules in finding the best solution and bound to the statistical structure of the diagnostic data [1]. However, the use of these methods for detection of defects does not provide quantitative individual characteristics of a serial product sample to assess the degradation, the development of defects and prognosis.

For information on the functioning of the GTE, it is primarily necessary to identify any stable diagnostic features with invariance and noise immunity to external influences. It is necessary to determine the diagnostic characteristics that accompany the operation of the facility over the life cycle, and provide information on specific device nodes. The diagnostic performance of symptom connected in a controlled setting, sensitivity, accuracy and invariance, as well as the possibility of implementing quantitative analysis, the use of classical methods of mathematical modeling of mechanical and electromechanical systems and forecasting. Basically, the amplitude method is used in the diagnosis of GTE to obtain the measurement information. In accordance with that defined by the controlled physical quantity, requirements of accuracy of measurements and mathematical processing. In the vibration (the amplitude method) is velocity, vibration acceleration and amplitude. These parameters carry information about all the interacting elements of the device, but sensitive to external influences and changes modes of operation. The problem is obtaining quantitative information about the degradation of specific parts of the mechanism. Also there is a background vibration characteristic of defect-free product and, therefore, much more complicated detection of incipient defects. Also used phase method, in which the monitored characteristics associated with the operational cycle of the device.

2.2. The requirements for diagnostic signs

The requirements for diagnostic symptom do not depend on a method of obtaining information, selecting the measured variables and formed by the next OS main conditions:

1. The measured physical quantity to the maximum extent depends on the interaction of structural elements and processes in the device, and to a much lesser extent from external influences and modes of operation.

2. Optimally, when the scan characteristic has the invariance and stability to various influences in the process of operation and to change modes of operation. The main quality characteristics during the entire life cycle, enabling the comparison and quantification of changes in the mechanism, should be preserved [2, 3].

3. The selection of the measured physical quantity and the accuracy of the transducers determine the possibility to quantify the slow processes that characterize the degradation mechanism, deterioration and identification of incipient defects. An error in the relative-ness of vibro-acoustic parameters from 0.1% to 10% corresponds to the level of gross errors. The measured physical quantity should be in the maximum degree does not depend on external influences, changes the modes, which is necessary for maximum repeatability of the diagnostic signs.

An example is the measurement of time intervals corresponding to the turnover and its shares, for a shaft turbine TPS at check 3 shutdowns of the generator of the turbine unit from an external network, showed repeatability of chronograms, providing analysis of the mode (Fig. 1) [6].

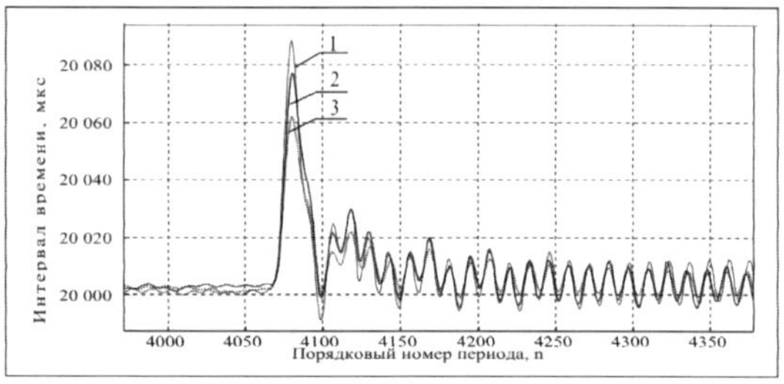

Fig. 1. The chronogram generator shut-down of turbine CHP plant in different modes from the external network

4. Should be provided with the quantitative correlation characteristics of the diagnostic signs of degradation of product design and processes in a functioning device. Due to problems binding the measurements to the design of mechanisms in technical diagnostics is dominated by peer review. With the use of vibroacoustic parameters, the main problem is the relationship of the controlled parameters (velocity and acceleration) solving a specific problem, for example, the definition of wear of the gear teeth in the gearbox, the quantitative wear of the bearing. An unsolved problem for large-sized products is to identify the sources of vibration.

5. Identify diagnostic signs should be extended to standard product without the need for adaptation to a particular series of condom sample.

Currently developed approaches to the definition of requirements for diagnostic systems, but remains unresolved the question of determining the estimation algorithm of operation of the device, detection of defects and prediction of their development. It is proposed that the use of statistical information on the work of the diagnostic mechanism will allow to assess to confidence level of the diagnosis of mechanism.

3 Metrological level

The metrological level of the measurement results should provide an opportunity to quantify changes in the mechanism in operation. The measured physical quantity and the accuracy of the transducers should be sufficient to record small changes of the controlled parameter characterizing the wear, degradation mechanism and incipient defects. To empower solutions to the above outlined problems for rotor systems proposes a phase-chronometric method of diagnostics of cyclic mechanisms [4–6], in which to perform the analysis of kinematic parameters of motion of the device in combination with mathematical modeling on the basis of the classical theory of machines and mechanisms to identify diagnostic signs of defects. The kinematic parameters of the working cycle of the machine, the nature of the interaction which persists throughout the life cycle, are controlled by using phase methods. Since the operating cycle has remained relatively stable in the functioning of the product, there are conditions to obtain stable diagnostic characteristics observed during the life cycle. For detail information and obtain the number-based estimates of changes in the mechanism provided the necessary metrological level (- relative error of 5•10-4 % at the industrial frequency of 50 Hz).

4 Mathematical modeling

The main feature of phase methods is the ability to use the classical theory of the description of machines and mechanisms, developed for a variety of cyclic mechanisms specialists in a specific area of technology to analyze the device and determine diagnostic features. The measurement information has a significantly smaller noise in comparison with the natural background vibrations of machines and mechanisms, what allows going to the level of incipient defects. The capabilities of computers allow the development of more detailed mathematical models for diagnostics and carrying out of computing experiment on revealing of diagnostic signs. The result of mathematical modeling is measuring information corresponding to experimental data, and there is a possibility of simulation of defects and classification of defects at the stage of creation, testing and production.

The mathematical model corresponding to the accurate measurements obtained by using high-precision measuring systems is a key factor of communication measurements with the processes of functioning of parts and components in the device. The detail of the results of computational experiment should correspond to the results of processing of experimental data obtained in precision measurements. For example, the comparison of mathematical modeling results and experimental data of functioning of the cutting machine, taken with a rotating spindle in the system drive-reducer-spidel presented.

5 The results of the comparison of mathematical simulation and experiment

An example of this approach is the creation of measurement monitoring of turbine TPS with the implementation of the diagnostic system and mathematical models of turbine T-250. The measuring system provided absolute error of measurement of time intervals corresponding to the turnover and its share in operating conditions $\pm 1 \cdot 10^{-7}$ s, which corresponds to turbine relative error of $5 \cdot 10^{-4}$ % of the nominal period on industrial-Thoth [2, 6, 7]. Spectral analysis is widely used to identify diagnostic features. In [8] describes the methods of recovery and interpolating the signals of the Fourier spectra. The spectrum parameters of torsional vibrations obtained by consecutively processing of the registered time intervals, sustainable, qualitatively and quantitatively change during the operation process, therefore suitable for the detection of diagnostic features.

Processing of the results of successive measurements of the respective revolution of the shaft and its share, built over a long period spectra of torsional vibrations of a shaft of the turbine (TA) T-250 [2] has preserved its major characteristics until now (Fig. 2).

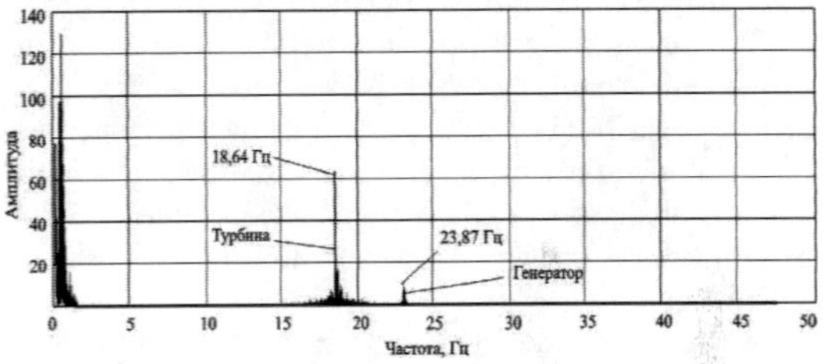

Fig.2. The range of torsional vibrations obtained from the experimental data for the turbine T-250

A mathematical model is developed to relate measurements of time intervals corresponding to a revolution of the shaft and the lobes, with structural TU elements. According to the results of the computational experiment produced a range of fundamental natural frequencies KRU-telnyh vibrations of a shaft (length 46 m, weight 130 tons, 3000 Rev/min) (Fig. 3).

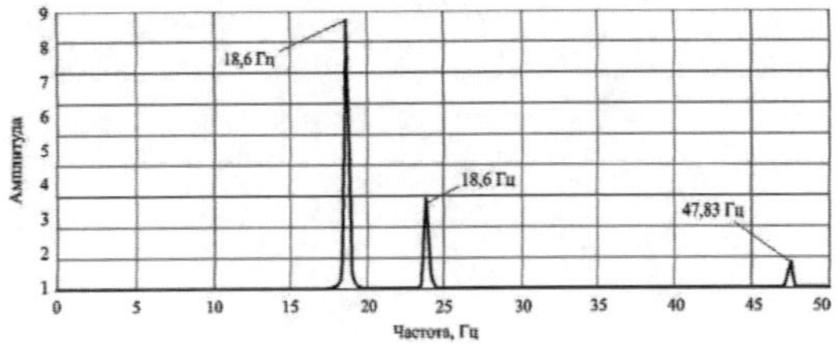

Fig. 3. The range of torsional vibrations obtained by mathematical simulation for the turbo unit T-250.

Comparison of results of mathematical modeling for fixed frequency 18,60 Hz and of 23.87 Hz with the results of experimental data processing of 18.64 Hz 23,20 Hz showed differences ranging from 0.2 to 2.9 %, which allows to model defects and shape classification of defects using mathematical models in a phase-chronometric performance.

The developed mathematical model [9] for turbine T-250 takes into account the relationship between the electrical parameters of the generator, geometrical and physical-mechanical parameters of the turbine, including the moments of inertia and rigidity of the connected sections of the shafting.

On the one hand, the mathematical model provides an interpretation of the experimental data and classification of defects at the stages of development, testing and production. On the other hand, the mathematical model allows taking into account the individual characteristics of the system turbine-generator (moments of inertia, stiffness, dimensional and mass characteristics, the electrical parameters of the generator). This provides the possibility of measuring and computing support electromechanical systems throughout the life cycle and determining the deviation of system parameters based on the identified diagnostic features. Part of a mathematical model THAT includes the description of the turbine and generator required for describing TU torsional vibrations, bending vibrations of a shaft of the turbine, parameters of Electromechanical processes occurring in the generator in the longitudinal and transverse axes rigidly associated with the rotor.

To perform work on the turbine T-200 in the mathematical model of TU T-250 was introduced, the original data is TU T-200 and pre-defined frequency of torsional vibrations of the shafting. After performing the experimental work was the comparison of the obtained experimental results with preliminary theoretical calculations.

According to the results of numerical experiments a spectra of the main natural frequencies of torsional vibrations of a shaft of the system generator-turbine was constructed. The experimental spectrum of frequencies of torsional vibrations of a shaft of TU T-200 obtained after further processing.

Comparison of results of mathematical modeling with the results of experimental data processing shows matching the frequency spectra obtained from periodogram of rotation of THE shaft and using a mathematical model.

The range has been pre-calculated using the mathematical model for TU T-250 and inputs TU T-200.

Its parameters differ from the experimental data for fundamental frequencies and 38,78 i.e. 45.98 Hz respectively from 0.39 % to 1.20 %, which testifies to the effectiveness of the phase approach for the analysis of cyclic mechanisms (including GTE) and identify diagnostic features.

6 Conclusion

Thus, the application of phase methods for assessing the current technical condition and emergency protection of aircraft engines will provide conditions to obtain a stable diagnostic criterion. In combination with mathematical modeling on the basis of the classical theory of machines and mechanisms and the high metrological level of realized correlation of measurement results in the design to solve specific problems and achieve incipient defects.

References

1. O.F. Mashoshin, *Diagnosis of aviation technology (information framework)* (Moscow, MGTUGA publ., 2007)
2. M.I. Kiselev, A.P. Kozlov, A.N. Morozov et al., *Izmeritel'naia tekhnika*, **12**, 28–29 (1996)
3. M.I. Kiselev, V.I. Proniakin, *Nauka i obrazovanie. MGTU im. N.E. Baumana*, **5** (2011) http://technomag.bmstu.ru/doc/192225.html
4. V.I. Pronyakin, *Measurement Techniques*, **51**(10), 1058–1064 (2008)
5. V.I. Proniakin, *Mir izmerenii* **9**, 57–61 (2011)
6. A.I. Baikov, M.I. Kiselev el al., *Gidrotekhnicheskoe stroitel'stvo*, **2**, 2–8 (2015)
7. M.I. Kiselev, V.I. Pronyakin, Ya.V. Chivilev, N.A. Zroichikov, *Thermal Engineering*, **53**(11), 868–872 (2006)
8. L.A. Aizenberg, B.A. Kravtsov, *Avtometriia*. **1**, 60–64 (1989)
9. A.S. Komshin, *Measurement Techniques*, **56**(8), 850–855 (2013)

Development of a Self-Stabilizing Robotic Chassis for Industry

Igor Ryadchikov[1], *Semyon* Sechenev[1,*], *Alexander* Svidlov[1], *Sergey* Sinitsa[1] and *Evgeny* Nikulchev[2,3]

[1]Kuban State University, 350040 Krasnodar, Russia
[2]Moscow Technological Institute, 119334 Moscow, Russia
[3]Moscow Technological Institute MIREA, 119454 Moscow, Russia

Abstract. Presented the description of the bipedal robotic chassis with the unique kinematic scheme which has the possibility to locomote in complicated multi-level environment. AnyWalker is equipped with the system of compensation of external impacts with motor-wheels which can self-stabilize the robotic system in 3 dimensions. Presented chassis suggests to have open software and hardware architecture in order to become the universal walking platform for service and industry robots.

1 Introduction

Service and personal robotics sets huge amount of tasks which require the ability of locomotion in complicated environment, sometimes multi-level with stairways, doors, windows, city obstacles. Sometimes there is no way to specially prepare the field of action i.e. make the surface flat or remove the obstacles. Therefore if the cost of environment preparing is too high there is the idea of creating the universal robotic system that is able to move in various surroundings.

The problems of creating anthropomorphic robots with the same purpose are discussed in [1]. There are no analytically solvable mathematical models of bipedal robots without independent stabilization systems. The task of creating the mechanical solution of anthropomorphic robot with the same degrees of freedom as human's is not solved yet.

It is proposed to incapsulate the bipedal locomotion function in complicates environment the robotic chassis with the possibility of hardware and software extension due to open architecture standarts.

2 Kinematic schemes

Bipodal robotic chassis is proposed, which has the ability to locomote in complicated not specially prepared environment and has the independent incapsulated built-in stabilization module.

The advantages of the kinematic scheme shown in Fig. 1:
- Large fiend of static stability zones;
- Low center of mass comparing to anthropomorphic design;
- The ability to roll on body and balance using segway principle and manipulate objects with two "hands";
- The ability to climb on the obstacle three times higher than the robot itself in a default position;
- The ability to stabilize standing on one "leg" and manipulating objects with the other;
- Better density comparing to the anthropomorphic construction having similar step size.

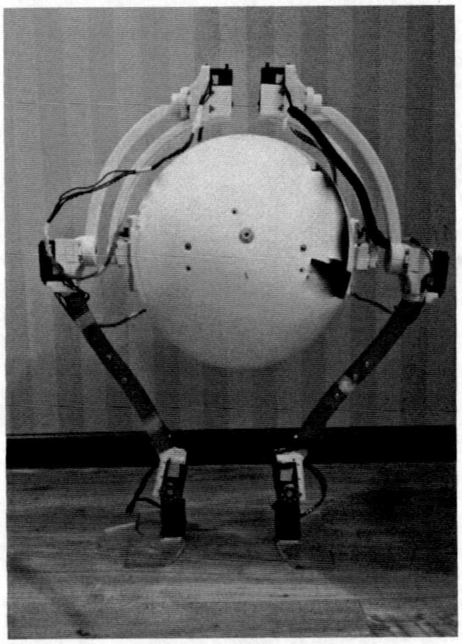

Fig. 1. General view of the prototype

Kinematic scheme is shown on the Fig. 2, where 1 – body of the robot, 2 – control unit, 4 – power supply, 5 – corrective center of mass displacement unit, 6 – surface, 7 – axis of the first and second hinges rotation, 8 – axis of gravity passing through the center of mass, 9 – obstacle, 11 – first part of the upper robot leg, 12 – second part of the upper robot leg, 13 – lower robot leg,

* Corresponding author: sechenev.semen@gmail.com

 DOI: 10.1051/matecconf/20179902007

14 – foot, 21 – first hinge, 22 – second hinge, 23 – third hinge, 24 – fourth hinge.

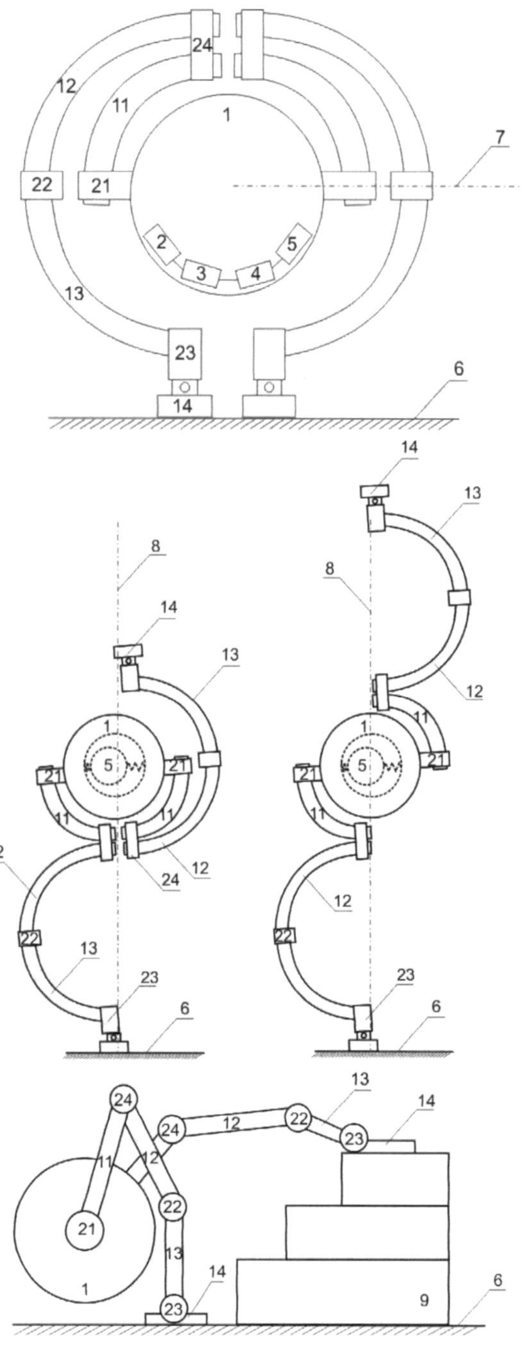

Fig. 2. Kinematic scheme

AnyWalker can reach continuous speed of 3 km/h and peak speed of 7,7 km/h with the form-factor of 0.3 m ball diameter while walking and 1.5 km/h while rolling.

With the board voltage of 16 V, servos average current assumption of 2.5 A, average speed of 0.8 m/s, stabilization system power of 2 kW and robot weight of 10 kg there is the cost of transportation equals to 1.85, where COT = energy / (weight * distance), which includes energy to run the motors and all electronics.

Comparisons:

- Toyota Prius COT = about 0.15
- Human COT = about 0.2 (a bit better than Ranger)
- Asimo COT = about 2 (54 kg, 1.5 m/s, 1.8 kW)
- BigDog COT = about 15 (12.8 miles, 4 gal gas, 110 kg)

3 Self-stabilization system

AnyWalker unique self-stabilization system consists of three motor-wheels, incapsulated built-in independent module, situated in orthogonal planes, and the centers of mass flywheels are the same. This stabilization system is inspired by the construction presented in [2-4].

While designing the AnyWalker chassis the problem of optimal parameters of the system selection is being faced. In particular, the task of deciding which flywheels to use in the design of the stabilization system, with the target optimization function of minimizing the weight of service structures, while maintaining the size of the field of management.

The problem of picking up a lighter wlywheel geometrically similar to original which will improve the quality of one-dimensional stabilization system with the parameters proposed in [4] has the following solution. The original flywheel has mass m_w = 0.204 kg and inertia moment $I_w = 0.57 \times 10^{-3}$ kg·m^2. Let k be the lineaar scaling factor of the flywheel, then the flywheel mass $m_w(k)$ = 0.204 k^3 kg and inertia moment $I_w = 0.57 \times 10^{-3} k^5$ kg·m^2. The maximum deviation angle α_{max} which the stabilization system can compensate dependence of k plot was build with the numeric methods described in [5]. It is seen that the $\alpha_{max}(k)$ function reaches maximum value at $k \approx 0.6$. The conclusion is that for the stabilization system with such parameters there is the optimal flywheel with mass m_w = 0.044 kg and inertia moment $I_w = 0.0443 \times 10^{-3}$ kg·m^2.

The following example (Fig. 3) shows the dependencies of the maximum deviation angle α_{max} of k with the different friction coefficients.

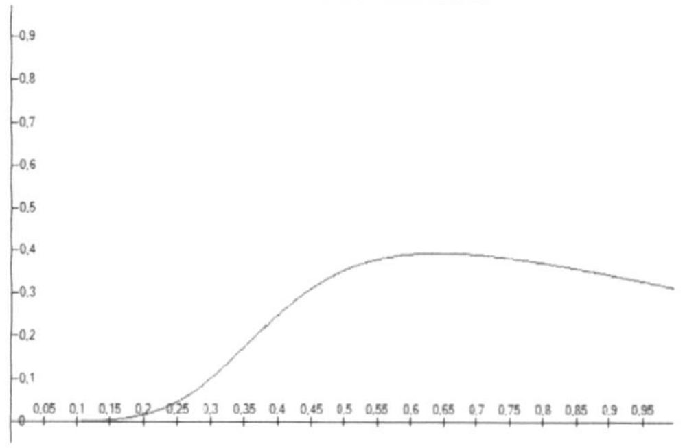

Fig. 3. The maximum deviation angle α_{max} which the stabilization system can compensate dependence of k plot

Fig 4 shows three modelling experiments, in which all of the parameters except the friction coefficients are the same. The green plot accords to the experiment friction coefficient *C*, blue – 10×*C*, red – 50×*C*. The explanation of the plots is that the higher friction coefficient gives more restrictions to changing the parameters of the system with keeping the stabilization of the analytical solution of differential equations system.

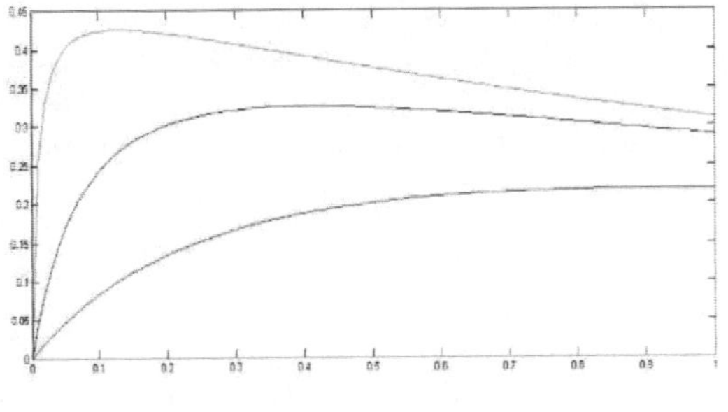

Fig. 4. Modelling experiments

Fig 5 shows the MATLAB simulation of the 2-axis stabilization with one «foot» hard connected to the ground.

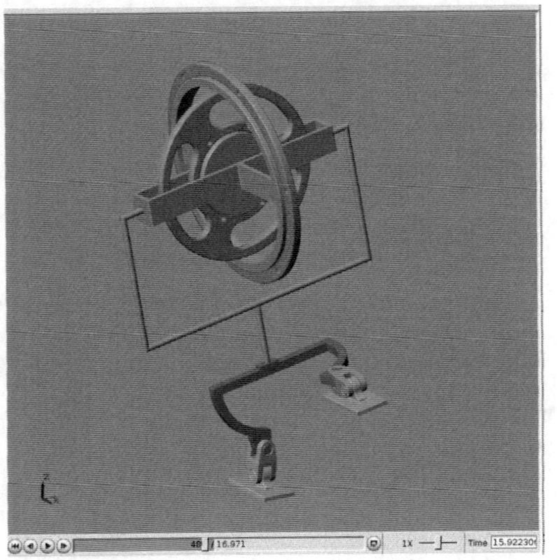

Fig. 5. Simulation of the 2-axis stabilization system in MATLAB

4 Hardware and software architecture of the control system

Hardware and software architecture of the control system AnyWalker consists of the following levels:

1. Reactive layer is based on a microcontroller STM32F429ZI. The lower reactive level performs robot stabilization using flywheels and robot movements using leg servos in a decoupled way. It receives the movement commands for the servos from Execution level and tries to run them keeping robot balance via flywheels, monitors servos temperature and power consumption. This level is implemented on the STM32F429 Discovery microcontroller. IMU6050 based accelerometers/gyroscopes are queried via I2C bus and filtered via Madgwick sensor fusion algorithm. Flywheels movement and speed computation are implemented using Maxon EPOS2 controllers with Maxon XXX motors. Dynamixel XXX servos for robot legs are controlled via RS485 using MAX485 based converter.

2. The Executive level is based on the controller TRIK. The Execution level is connected to Reactive level via UART or Bluetooth. We created the mobile application for Android platform which sends movement commands to AnyWalker and displays the result. TRIK and ROS variants are planned.

3. Algorithmic level represents the application software installed on a personal computer or smartphone, completely or partially in the cloud. The Application level is planned to provide high level user interface, implement cloud services client API to recognize voice commands, gather sensor and commands data for machine learning purposes, connect several AnyWalkers with patterns of crowd behaviour.

It is proposed to use the control algorithmic support as an information service that will allow you to give third-party API researchers to solve applied problems.

The levels of the software architecture provide an API to the next level according to their functionality [5]:

1. Reactive level: the RPC API;
2. Executive level: C++ API, API QtSrcipt;
3. Algorithmic level: the SOAP API.

AnyWalker chassis is presented to be a chassis for applied robotics with the following expansion possibilities:

Hardware level: adding different sensors, changing manipulators, increasing or decreasing the robot size. Provides the ability for third-party engineers to develop and sell their own hardware that is compatible with AnyWalker;

Control level: programming new movements, specialization, programming new manipulators, adding degrees of freedom.

Intelligent systems level: designing crowd AnyWalker patterns, organizing robotic competitions, artificial intellect systems on the basis of the chassis. Provides the ability for developers to create and implement their own applications.

5 Conclusions

Proposed bipodal robotic chassis is able to locomote in complicated not specially prepared environment and has the independent incapsulated built-in stabilization module and non-anthropomorphic design.

Proposed kinematic scheme allows to have cost of transportation around 2 with the diameter of robot body of 0.3 m and robot weight of around 10 kg, which is similar to the CoT of the Asimo robot.

Proposed optimization methods and criteria of the designed systems with stabilization based on the principles of the inverted pendulum with flywheel. Proposed stabilization system parameters calculation method allows minimizing the weight of service structures thus increasing the robotic system payload. Proposed scheme of open software and hardware architecture of the platform provides the possibility to create extensions and build various applied robotic systems based on AnyWalker locomotion module.

References

1. N.A. Chaturvedi, A.M. Bloch, N.H. McClamroch. *Proceedings of American Control Conference*, 3612-3617 (2006)
2. M. Muehlebach, G. Mohanarajah, R. D'Andrea, *2013 IEEE 52nd Annual Conference on Decision and Control (CDC)*, 1283-1288 (2013)
3. M. Gajamohan, M. Muehlebach, T. Widmer, R. D'Andrea. *2013 European Control Conference (ECC)*, 268-274 (2013)
4. M. Gajamohan, M. Merz, I. Thommen, and R. D'Andrea, *2012 IEEE/RSJ International Conference on Intelligent Robots and Systems (IROS)*, 3722-3727 (2012)
5. I. Riadchykov, S. Sechenev, S. Sinitsa, E. Nikulchev, *6th Seminar on Industrial Control Systems: Analysis, Modeling and Computation*, **6**, 01003 (2016)

Modeling the Motion of a Flapping Wing Aerial Vehicle

L.Y. Vorochaeva[1,*], *S.V.* Efimov[1], *O.G.* Loktionova[1], *B.V.* Lushnikov[1] and *S.F.* Jatsun[1]

[1]Southwest State University, 305040 Kursk, Russia

Abstract. The article discusses the vertical flight of a flapping wing aerial vehicle, which is also called an ornithopter. The robot is a chain of five links connected in series by active cylindrical hinges with the central link being the body and the remainder forming folding wings in pairs. The distinctive feature of this device is that the flaps of its wings imitate those of a seagull i.e. the device has a biological prototype. We construct a mathematical model of this device; much attention is given to the model of the interaction of the wings with the air environment and we determine the positions and velocities of points of application of the reduced aerodynamic forces to each of the links. Based on the results of numerical modelling of the vertical flight of the robot three modes of flight were established: ascent, hovering at a certain height and descent. The device can operate in these modes based on the oscillation parameters of the wings in particular flapping frequency and amplitude, the ratio of the amplitudes of two links and one wing and the shift of the equilibrium oscillation position of the wings relative to zero.

1 Introduction

With the development of robotics and related technologies more and more interest is shown towards bionics - the science that studies the nature of the movement of living organisms, as well as the phenomena and processes taking place in them in order to create mechatronic devices operating on the same principles.

The study of the character of the movement of birds has enabled us to makes an important conclusion that a system of three or five rigid bodies connected by hinges equipped with electric drives can be used as a simplified system that adequately models the motion of a robot-ornithopter [1-7]. By feeding control voltages to the actuators we can change the relative angles defining the position of the links with respect to each other by given values.

2 Description of the ornithopter

The article discusses a flapping wing device called an ornithopter, which consists of a chain of five links [5]. Each wing of the robot is formed by two links: 1 and 2, 4 and 5, connected to each other by cylindrical hinges that also connect them with the body 3 (fig. 1). The cylindrical hinges are active due to actuators 6 that create the flapping motion of the wings relative to the body. The distinctive feature of this work is that the horizontal projection of the area of the wings of the device changes in each flap, which results in flight: the wings fold during the upward flap causing a reduction in aerodynamic resistance on the wings and unfold during the downward flap causing an increase in the in aerodynamic resistance under the wings.

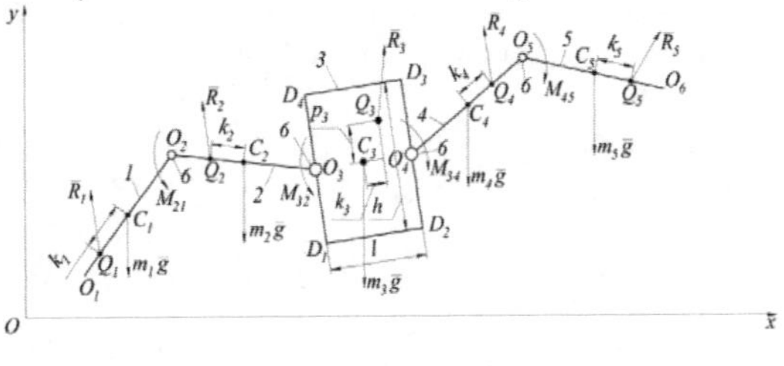

Fig. 1. Settlement scheme of the ornithopter.

The work studies the motion of an ornithopter in the vertical plane Oxy under the condition that the body of the device is modeled by a rectangle with sides l_3 and h_3 and the links – by rods of length l_i. We also assume that the centres of mass of the links coincide with their symmetry points - C_i. The relative coordinate system $O_ix_iy_i$ is associated with each link such that the axis O_ix_i is located along the link (for link 3 the relative coordinate system $C_3x_3y_3$ the axis C_3x_3 passes through the points C_3 and O_4). The position of the robot in the Oxy plane is uniquely described by the projections of the position of the body's center of mass on axes Ox and Oy and the angles of inclination of the links to the horizontal plane φ_i, which are measured in the anticlockwise direction. Thus the general coordinates vector has the form

$$\bar{q} = \left(x_{C3} \quad y_{C3} \quad \varphi_1 \quad \varphi_2 \quad \varphi_3 \quad \varphi_4 \quad \varphi_5\right)^T \quad (1)$$

The flight of the ornithopter takes place under the action of gravitational forces, m_ig acting at the centers of

[*] Corresponding author: mila180888@yandex.ru

mass of the links, moments M_{21}, M_{32}, M_{34} and M_{45} generated by the actuators and aerodynamic forces R_i applied at points Q_i.

3 Aerodynamic model

3.1. Determination of aerodynamic forces

During the motion of the ornithopter aerodynamic forces act on every link i=1-5. These forces can be written as components on the absolute and relative coordinate axes as follows (fig. 2):

$$\overline{R}_i = \overline{R}_{ix} + \overline{R}_{iy} = \overline{R}_{ix}^{(i)} + \overline{R}_{iy}^{(i)} \quad (2)$$

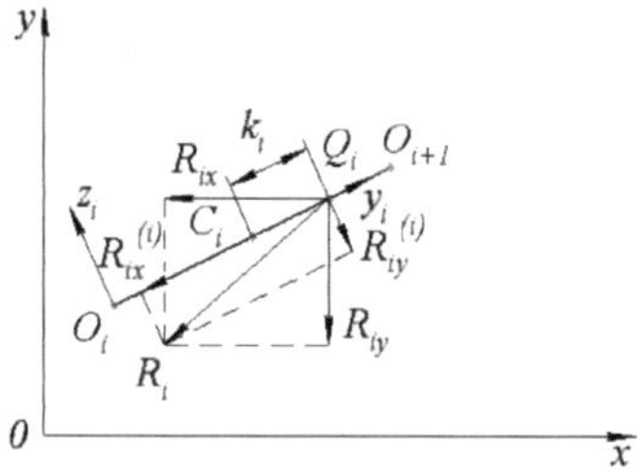

Fig. 2. Settlement scheme of aerodynamic forces on links i=1, 2, 4, 5.

The component of aerodynamic force in the absolute coordinate system is given by the following formulas:

$$R_{iz} = \frac{C_x \rho S_i^{Ox}}{2} \dot{x}_{Qi} \left|\dot{x}_{Qi}\right|, \quad R_{iy} = \frac{C_y \rho S_i^{Oy}}{2} \dot{y}_{Qi} \left|\dot{y}_{Qi}\right|, \quad (3)$$

where C_x, C_y - coefficients of the aerodynamic force components, S_i^{Ox} and S_i^{Oy} - areas of sides of i-th link perpendicular to axes Ox and Oy respectively, $\dot{x}_{Qi}$, $\dot{y}_{Qi}$ - velocity components of points Q_i on axes Ox and Oy.

We calculate areas S_i^{Ox} and S_i^{Oy} by the formulas:

$$S_i^{Ox} = S_i^{Oixi} \cos\varphi_i + S_i^{Oiyi} \sin\varphi_i \quad (4)$$

$$S_i^{Oy} = S_i^{Oixi} \sin\varphi_i + S_i^{Oiyi} \cos\varphi_i \quad (5)$$

where

$$S_i^{Oixi} = b_i h_i, \; S_i^{Oiyi} = l_i b_i \quad (6)$$

- areas of the links peprendicular to axes $O_i x_i$ and $O_i y_i$ of the relaive coordinate systems for i=1, 2, 4, 5 $S_i^{Oixi} = 0$, b_i – the width of the links (not shown in fig. 1) determined in the direction perpendicular to plane Oxy.

In order to determine the positions of points Q_i on the links we assume that points Q_i are the centers of mass of the figure formed by forces $\overline{R}_{iy}^{(i)}$ acting peperndicular t the links. The links are located at distances of k_i from the centers of mass of the links. The position of point Q_3 is determined by the destances, k_3 and p_3 relative to the center of mass of the latter along sides D_1D_2 and D_2D_3.

3.2 Determination of the velocities of the points Q_i

The velocities of the points Q_i are determined from their positions on the links given earlier, which constantly change during the robot's motion. The velocities of points Q_2 and Q_4 equal:

$$\dot{\vec{r}}_{Q2} = \dot{\vec{r}}_{C3} + \dot{T}_{30}\vec{\rho}_{C3O3}^{(3)} + \dot{T}_{20}\left(\vec{\rho}_{O3C2}^{(2)} + \vec{\rho}_{C2Q2}^{(2)}\right), \quad (7)$$

$$\dot{\vec{r}}_{Q4} = \dot{\vec{r}}_{C3} + \dot{T}_{30}\vec{\rho}_{C3O4}^{(3)} + \dot{T}_{40}\left(\vec{\rho}_{O4C4}^{(4)} + \vec{\rho}_{C4Q4}^{(4)}\right). \quad (8)$$

where $\vec{\rho}_{C2Q2}^{(2)} = (k_2 \;\; 0)^T$, $\vec{\rho}_{C4Q4}^{(4)} = (k_4 \;\; 0)^T$ - the corresponding relative radius-vectors, $\dot{T}_{i0}$ - the derivative of the rotation matrix

$$T_{i0} = \begin{pmatrix} \cos\varphi_i & -\sin\varphi_i \\ \sin\varphi_i & \cos\varphi_i \end{pmatrix} \quad (9)$$

The velocities of points Q_1, Q_5 equal:

$$\dot{\vec{r}}_{Q1} = \dot{\vec{r}}_{C3} + \dot{T}_{30}\vec{\rho}_{C3O3}^{(3)} + \dot{T}_{20}\vec{\rho}_{O3O2}^{(2)} + \dot{T}_{10}\left(\vec{\rho}_{O2C1}^{(1)} + \vec{\rho}_{C1Q1}^{(1)}\right) \quad (10)$$

$$\dot{\vec{r}}_{Q5} = \dot{\vec{r}}_{C3} + \dot{T}_{30}\vec{\rho}_{C3O4}^{(3)} + \dot{T}_{40}\vec{\rho}_{O4O5}^{(4)} + \dot{T}_{50}\left(\vec{\rho}_{O5C5}^{(5)} + \vec{\rho}_{C5Q5}^{(5)}\right) \quad (11)$$

where $\vec{\rho}_{C1Q1}^{(2)} = (k_1 \;\; 0)^T$, $\vec{\rho}_{C5Q5}^{(5)} = (k_5 \;\; 0)^T$ - the corresponding relative radius-vectors.

The velocity of point Q_3 has the form:

$$\dot{\vec{r}}_{Q3} = \dot{\vec{r}}_{C3} + \dot{T}_{30}\vec{\rho}_{C3Q3}^{(3)} \quad (12)$$

where $\vec{\rho}_{C3Q3}^{(3)} = (k_3 \;\; p_3)^T$ - the relative radius-vector.

3.3 Determination of the positions of points Q_i

In order to determine the locations of points Q_i on the links (distances k_i and p_3) we consider the corresponding aerodynamic force components $R_{iy}^{(i)}$ and $R_{3x}^{(3)}$, perpendicular to axes $O_i x_i$ and $O_3 y_3$ and assume points Q_i to be the cemters of mass of the corresponding figures. The velocity components of points Q_i in the $O_i x_i y_i$ cordinate system are writtend using the rotation matrix T_{0i}:

$$\dot{r}_{Qi}^{(i)} = T_{0i}\dot{r}_{Qi} \quad (13)$$

$$T_{0i} = T_{i0}^{-1} = \begin{pmatrix} \cos\varphi_i & \sin\varphi_i \\ -\sin\varphi_i & \cos\varphi_i \end{pmatrix}. \quad (14)$$

DOI: 10.1051/matecconf/20179902008

We move on to evaluate the distances k_i. For this purpose we determine the formulas for $\left(\dot{y}_{Qi}^{(i)}\right)^2$ by grouping the terms before k_i^2, k_i, k_i^0 respectively and obtain formulars for calculating the components of aerodynamic forces on axes O_iy_i. We present $R_{iy}^{(i)}$ as follows

$$R_{iy}^{(i)} = a_{iy}k_i^2 + b_{iy}k_i + c_{iy}\,. \qquad (15)$$

We introduce the denotations below.

$$a_{iy} = \frac{C_y \rho S_i^{Oiyi}}{2}\dot{\varphi}_i^2 \qquad (16)$$

$$\begin{aligned} b_{i=1,5y} = {} & \frac{C_y \rho S_i^{Oiyi}}{2}[-2\dot{x}_{C3}\dot{\varphi}_i \sin\varphi_i \\ & + 2\dot{y}_{C3}\dot{\varphi}_i \cos\varphi_i \mp \dot{\varphi}_i\dot{\varphi}_3 l_3 \cos(\varphi_3 - \varphi_i) \mp \\ & \mp 2\dot{\varphi}_i\dot{\varphi}_{2(4)} l_{2(4)} \cos(\varphi_{2(4)} - \varphi_i) \mp \dot{\varphi}_i^2 l_i\,] \end{aligned} \qquad (17)$$

$$\begin{aligned} b_{i=2,4y} = {} & \frac{C_y \rho S_i^{Oiyi}}{2}\left[-2\dot{x}_{C3}\dot{\varphi}_i \sin\varphi_i + \right. \\ & \left. + 2\dot{y}_{C3}\dot{\varphi}_i \cos\varphi_i \mp \dot{\varphi}_i\dot{\varphi}_3 l_3 \cos(\varphi_3 - \varphi_i) \mp \dot{\varphi}_i^2 l_i\right] \end{aligned} \qquad (18)$$

$$b_{3y} = -C_y \rho S_3^{O3y3}\dot{\varphi}_3\left(\dot{y}_{C3}\cos\varphi_3 - \dot{x}_{C3}\sin\varphi_3\right) \qquad (19)$$

$$\begin{aligned} c_{i=1,5y} = {} & \frac{C_y \rho S_i^{Oiyi}}{2}[\dot{y}_{C3}^2 \sin^2\varphi_i - \\ & 2\dot{x}_{C3}\dot{y}_{C3}\sin\varphi_i\cos\varphi_i + \dot{y}_{C3}^2\cos^2\varphi_i \pm \\ & \pm \dot{x}_{C3}\dot{\varphi}_3 l_3 \sin\varphi_i \cos(\varphi_3 - \varphi_{i1}) \pm \\ & 2\dot{x}_{C3}\dot{\varphi}_{i\pm1} l_{i\pm1}\sin\varphi_i\cos(\varphi_{i\pm1} - \varphi_i) \pm \dot{x}_{C3}\dot{\varphi}_i l_i \sin\varphi_i \mp \\ & \mp \dot{y}_{C3}\dot{\varphi}_3 l_3 \cos\varphi_i \cos(\varphi_3 - \varphi_i) \mp \\ & 2\dot{y}_{C3}\dot{\varphi}_{i\pm1} l_{i\pm1}\cos\varphi_i\cos(\varphi_{i\pm1} - \varphi_i) \mp \dot{y}_{C3}\dot{\varphi}_i l_i \cos\varphi_i + \\ & + \dot{\varphi}_3^2 l_3^2 \cos^2(\varphi_3 - \varphi_i)/4 + \\ & \dot{\varphi}_{i\pm1}\dot{\varphi}_3 l_{i\pm1} l_3 \cos(\varphi_{i\pm1} - \varphi_i)\cos(\varphi_3 - \varphi_i) + \\ & \dot{\varphi}_i\dot{\varphi}_3 l_i l_3 \cos(\varphi_3 - \varphi_i)/2 + \\ & + \dot{\varphi}_{i\pm1}^2 l_{i\pm1}^2 \cos^2(\varphi_{i\pm1} - \varphi_i) + \\ & \left. + \dot{\varphi}_i\dot{\varphi}_{i\pm1} l_i l_{i\pm1}\cos(\varphi_{i\pm1} - \varphi_i) + \dot{\varphi}_i^2 \frac{l_i^2}{4}\right] \end{aligned} \qquad (20)$$

$$\begin{aligned} c_{i=2,4y} = {} & \frac{C_y \rho S_i^{Oiyi}}{2}[\dot{x}_{C3}^2 \sin^2\varphi_i - 2\dot{x}_{C3}\dot{y}_{C3}\sin\varphi_i\cos\varphi_i + \\ & + \dot{y}_{C3}^2\cos^2\varphi_i \pm \dot{x}_{C3}\dot{\varphi}_3 l_3 \sin\varphi_{i+1}\cos(\varphi_3 - \varphi_i) \pm \\ & \dot{x}_{C3}\dot{\varphi}_i l_i \sin\varphi_i \mp \dot{y}_{C3}\dot{\varphi}_3 l_3 \cos\varphi_i\cos(\varphi_3 - \varphi_i) \mp \dot{y}_{C3}\dot{\varphi}_i l_i \cos\varphi_i + \\ & \left. + \dot{\varphi}_3^2 \frac{l_3^2}{4}\cos^2(\varphi_3 - \varphi_i) + \dot{\varphi}_i\dot{\varphi}_3 l_i l_3 \cos(\varphi_3 - \varphi_i)/2 + \dot{\varphi}_i^2 \frac{l_i^2}{4}\right] \end{aligned} \qquad (21)$$

$$\begin{aligned} c_{3y} = {} & \frac{C_y \rho S_3^{O3y3}}{2}\left[\dot{x}_{C3}^2 \sin^2\varphi_3 \right. \\ & \left. - 2\dot{x}_{C3}\dot{y}_{C3}\sin\varphi_3\cos\varphi_3 + \dot{y}_{C3}^2\cos^2\varphi_3\right] \end{aligned} \qquad (22)$$

We find the distances k_i, i=1-5 by the formulas

$$k_i = \frac{\int_{-li/2}^{li/2} k_i \cdot R_{iy}^{(i)} dk_i}{\int_{-li/2}^{li/2} R_{iy}^{(i)} dk_i} = \frac{b_{iy} l_i^2}{a_{iy} l_i^2 + 12c_{iy}} \qquad (23)$$

Thus, the locations of points Q_i for the wings (links i=1,2,4,5) are found.

We calculate the distance $p_3$in the same way : we determine the formula for evaluating $\left(\dot{x}_{Q3}^{(3)}\right)^2$, transform the expression $\left(\dot{x}_{Q3}^{(3)}\right)^2$ by grouping the terms before p_3^2, p_3, p_3^0 and susbstituting the expression $\left(\dot{x}_{Q3}^{(3)}\right)^2$ to obtain the formula for determining the aerodynamic force component on axis O_3x_3.

We introduce the following denotations:

$$a_{3x} = \frac{C_x \rho S_3^{O3x3}}{2}\dot{\varphi}_3^2\,, \qquad (24)$$

$$b_{3x} = -C_x \rho S_3^{O3x3}\dot{\varphi}_3\left(\dot{x}_{C3}\cos\varphi_3 + \dot{y}_{C3}\sin\varphi_3\right) \qquad (25)$$

$$\begin{aligned} c_{3x} = {} & \frac{C_x \rho S_3^{O3x3}}{2}\left[\dot{x}_{C3}^2 \cos^2\varphi_3 \right. \\ & \left. + 2\dot{x}_{C3}\dot{y}_{C3}\sin\varphi_3\cos\varphi_3 + \dot{y}_{C3}^2\sin^2\varphi_3\right] \end{aligned} \qquad (26)$$

and write the force $R_{3x}^{(3)}$ as follows

$$R_{3x}^{(3)} = a_{3x}p_3^2 + b_{3x}p + c_{3x} \qquad (27)$$

We find the distance p_3 according to the formula

$$p_3 = \frac{\int_{-h3/2}^{h3/2} p_3 \cdot R_{3x}^{(3)} dp_3}{\int_{-h3/2}^{h./2} R_{3x}^{(3)} dp_3} = \frac{b_{3x} h_3^2}{a_{3x} h_3^2 + 12c_{3x}} \qquad (28)$$

For the special case when the velocity of rotation, $\dot{\varphi}_i \to 0$, the expression for the application points of the aerodynamic force will have the form:

$$a_{3x}(\dot{\varphi}_3 \approx 0) = 0\,,\ a_{iy}(\dot{\varphi}_i \approx 0) = 0\,,\ i\text{=1-5}, \qquad (29)$$

$$b_{3x}(\dot{\varphi}_3 \approx 0) = 0\,,\ b_{iy}(\dot{\varphi}_i \approx 0) = 0\,,\ i\text{=1-5}, \qquad (30)$$

$$c_{3x}(\dot{\varphi}_3 \approx 0) \neq 0\,,\ c_{iy}(\dot{\varphi}_i \approx 0) \neq 0\,,\ i\text{=1-5}, \qquad (31)$$

$$p_3 = 0\,,\ k_i = 0\,. \qquad (32)$$

Therefore we can conclude that for $\dot{\varphi}_i \approx 0$ the points of application of the aerodynamic force on links i=1-5 will be located at their centers of mass.

4 Mathematical Model for the Flight of the Ornithopter

The system of differential equations of the ornithopter's motion written using the 2nd order Lagrange equation and the virtual work principle can be presented as follows:

$$A(q)\ddot{q} + B(q)D(\dot{q})\dot{q} = F \tag{33}$$

where $A(q)$, $B(q)$ are matrices of coefficients, $D(\dot{q})$ – the diagonal matrix of the first derivatives of the general coordinates q and F – the matrix of generalised forces

$$A(q) = \begin{pmatrix} a_{11} & 0 & a_{13} & a_{14} & a_{15} & a_{16} & a_{17} \\ 0 & a_{22} & a_{23} & a_{24} & a_{25} & a_{26} & a_{27} \\ a_{31} & a_{32} & a_{33} & a_{34} & a_{35} & 0 & 0 \\ a_{41} & a_{42} & a_{43} & a_{44} & a_{45} & 0 & 0 \\ a_{51} & a_{52} & a_{53} & a_{54} & a_{55} & a_{56} & a_{57} \\ a_{61} & a_{62} & 0 & 0 & a_{65} & a_{66} & a_{67} \\ a_{71} & a_{72} & 0 & 0 & a_{75} & a_{76} & a_{77} \end{pmatrix}, \tag{34}$$

$$B(q) = \begin{pmatrix} 0 & 0 & b_{13} & b_{14} & b_{15} & b_{16} & b_{17} \\ 0 & 0 & b_{23} & b_{24} & b_{25} & b_{26} & b_{27} \\ 0 & 0 & 0 & b_{34} & b_{35} & 0 & 0 \\ 0 & 0 & b_{43} & 0 & b_{45} & 0 & 0 \\ 0 & 0 & b_{53} & b_{54} & 0 & b_{56} & b_{57} \\ 0 & 0 & 0 & 0 & b_{65} & 0 & b_{67} \\ 0 & 0 & 0 & 0 & b_{75} & b_{76} & 0 \end{pmatrix}, \tag{35}$$

$$D(\dot{q}) = \begin{pmatrix} \dot{q}_1 & 0 & 0 & 0 & 0 & 0 & 0 \\ 0 & \dot{q}_2 & 0 & 0 & 0 & 0 & 0 \\ 0 & 0 & \dot{q}_3 & 0 & 0 & 0 & 0 \\ 0 & 0 & 0 & \dot{q}_4 & 0 & 0 & 0 \\ 0 & 0 & 0 & 0 & \dot{q}_5 & 0 & 0 \\ 0 & 0 & 0 & 0 & 0 & \dot{q}_6 & 0 \\ 0 & 0 & 0 & 0 & 0 & 0 & \dot{q}_7 \end{pmatrix}, \tag{36}$$

$$F = (F_1 \;\; F_2 \;\; F_3 \;\; F_4 \;\; F_5 \;\; F_6 \;\; F_7)^T \tag{37}$$

5 Modeling of Vertical Flight of the Ornithopter

The analytical diagram of the ornithopter proposed in this work allows the device to imitate the flapping motion of birds and consequently, their behaviour during flight. In this connection, we constructed the laws of variation of the angles of the wing links, for which the behavior of the device is analogous to that of seagull [5].

$$\varphi_2(t) = -\varphi_{02}\cos(\omega t) - w_2 \tag{38}$$

$$\varphi_4(t) = \varphi_{04}\cos(\omega t) + w_4 \tag{39}$$

$$\varphi_1(t) = \begin{cases} \varphi_{01}\cos(\omega_1 t) + w_1, \; t < \dfrac{T}{4} \\ -\varphi_{01}\cos\left(\omega\left(t - \dfrac{T}{4}\right)\right) + w_1, \; \dfrac{T}{4} \le t < \dfrac{3T}{4} \\ \varphi_{01} + w_1, \dfrac{3T}{4} \le t \le T \end{cases} \tag{40}$$

$$\varphi_5(t) = \begin{cases} -\varphi_{05}\cos(\omega_1 t) - w_5, \; t < \dfrac{T}{4} \\ \varphi_{05}\cos\left(\omega\left(t - \dfrac{T}{4}\right)\right) - w_5, \; \dfrac{T}{4} \le t < \dfrac{3T}{4} \\ -\varphi_{05} - w_5, \dfrac{3T}{4} \le t \le T \end{cases} \tag{41}$$

where φ_{0i} is a certain value of the rotation angle of the link that determines the amplitude of the oscillations, w_i is the displacement of the equillibrium position of the link's oscillation upwards/downwards, T - the period of oscillations of the link. We have $\varphi_{05}^{-} = -\varphi_{05} - w_5$, $\varphi_{05}^{+} = \varphi_{05} - w_5$, $\varphi_{04}^{-} = -\varphi_{04} + w_4$, $\varphi_{04}^{+} = \varphi_{04} + w_4$.

We introduce the following denotations: $\varphi_{01} = \varphi_{05} = \varphi_0^{'}$, $\varphi_{02} = \varphi_{04} = \varphi_0$, $w_1 = w_2 = w_4 = w_5 = w$ and assume that $\omega_1 = 2\omega$. We introduce a coefficient characterising the ration of the oscillation amplitudes of links 4 and 5 (1 and 2)

$$\lambda = \frac{\varphi_{04}}{\varphi_{05}} = \frac{\varphi_0}{\varphi_0^{'}}. \tag{42}$$

The time graphs of the coordinate of the body's center of mass, y_{C3} for different values of the displacement ratios w_4=-w_5 are shown in fig. 3. It can be seen from the graphs that there are three modes of vertical flight of the device: ascent, hovering and descent. As the ratios w_4=-w_5 increase the descent mode switches to hovering, then ascent. For negative values of w_4=-w_5 the distance descended by the ornithopter decreases as the ratios approach 0. The distance by which the device ascends also decreases for positive values of w_4=-w_5 as the ratios approach 0|.

It is more convenient to analyse the dependence of the vertical speed of the device on the flapping amplitude of the wings for different values of oscillation parameters. The average speeds of motion of the body's centre of mass, $\dot{y}_{sr}$ for different modes of motion are presented in figs. 4-6. The average speed is given by the formula

$$\dot{y}_{sr} = \frac{1}{N}\sum_{j=1-N} \dot{y}_{C3},$$

DOI: 10.1051/matecconf/20179902008

where N is the number of calculations for one period of wing oscillation.

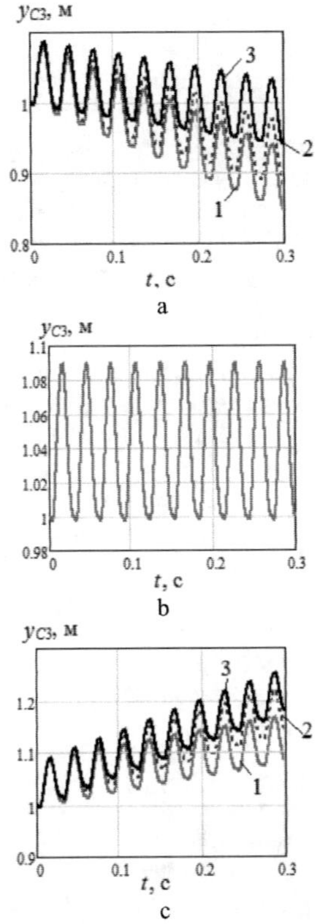

Fig. 3. Graphs of $y_{C3}(t)$ for T=0.03 s, φ_0=0.4 rad: a - descent (1 - w_4=-w_5=-15^0, 2 - w_4=-w_5=-10^0, 3 - w_4=-w_5=-5^0), b - hovering (w_4=-w_5=-1^0), c - ascent (1 - w_4=-w_5=5^0, 2 - w_4=-w_5=10^0, 3 - w_4=-w_5=15^0).

It can be seen from the graphs in fig. 4 that as the frequency, ω decreases (increase in the period of oscillations, T) the values of the average speeds decrease and a maximum is observed for all curves, which is evened-out as ω decreases.

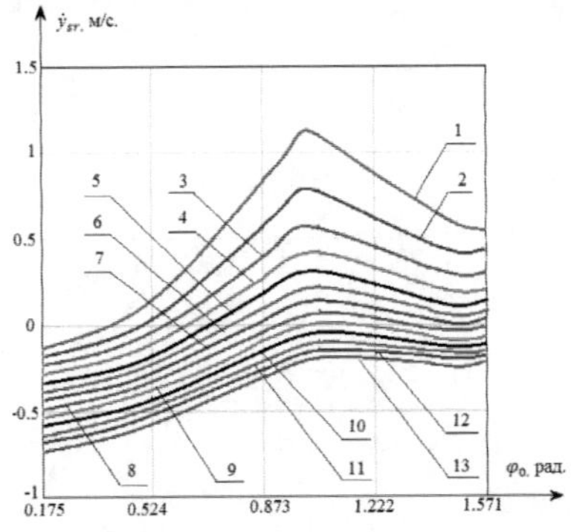

Fig. 4. Graphs of : 1 - T=0,03 s, 2 - T=0,04 s, 3 - T=0,05 s, 4 - T=0,06 s, 5 - T=0,07 s, 6 - T=0,08 s, 7 - T=0,09 s, 8 - T=0,10 s, 9 - T=0,11 s, 10 - T=0,12 s, 11 - T=0,13 s, 12 - T=0,14 s, 13 - T=0,15 s.

At the smallest angle, φ_0=0.175 rad the values of the average velocities are negative and at the greatest angle, φ_0=1.571 – negative for small oscillation frequencies and positive for large values of ω. This means that for large frequencies the following alternation of the modes of motion occurs depening on the increase in the amplitude of oscillations: descent, hovering and ascent. However, when the frequencies decrease the alternation sequence of the modes of motion is different: descent, hovering, ascent, hovering, descent and there is only one mode of motion at the smallest frequencies.

Fig. 5 shows the results of the study of the influence of the amplitude ratio of the oscillation amplitude of the links on the average values of the velocity of vertical displacement of the device.

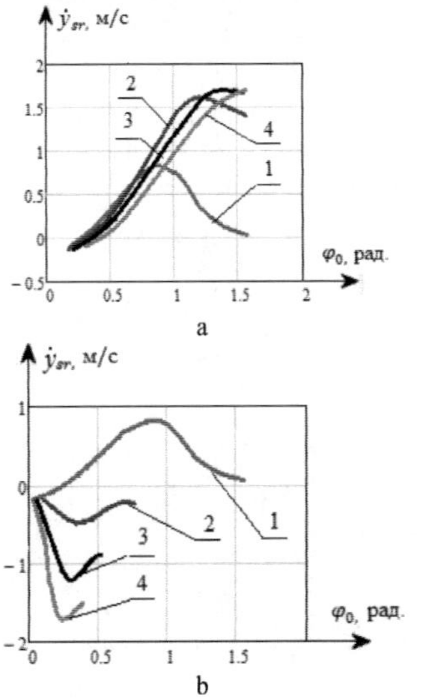

Fig. 5. Graphs of $\dot{y}_{sr}(\varphi_0)$ for w=0 rad, T=0,03 s: a - 1 - λ=1, 2 - λ=2, 3 - λ=3, 4 - λ=4, b - 1 - λ=1, 2 - λ=1/2, 3 - λ=1/3, 4 - λ=1/4.

We can single out three types of curves from the given graphs. The first type holds for the amplitude λ=1 when the average velocity $\dot{y}_{sr}$ is negative and close to zero at the minimum value of φ_0. As φ_0 increases up to the value, φ_0=0.785 rad the average velocity, $\dot{y}_{sr}$ passes through zero and attains its greatest value on a certain curve and then decreases by a curvilinear law and approaches zero at φ_0=1.571. the second type of graphs corresponds to the case when λ>1. In this case the greatest value of average velocity and the value of φ_0 corresponding to the maximum incerase, while the second part in which $\dot{y}_{sr}$ decreases is reduced and ceases to exist when λ=4. The third type of curves corresponds to the case when λ<1. In this case the average velocities only have negative values. Take note that as φ_0 increases these values smoothly decrease to miimum values and then smoothly increase and as λ decreases the peak of the average velocity if observed at the smallest value of φ_0, while the section of the graph corresponding to the increase of $\dot{y}_{sr}$ is reduced.

Fig. 6 shows the graphs of average vertical velocities of the ornithopter for different values of w. Three sections can be singled out on these graphs, which differ for positive and negative values of w. In the first case (for w<0) the average velocity at first increases almost proportionally to the amplitude of the links' oscillations, φ_0, attains the greatest value, then decreases also proportionally to the increase in φ_0, attais the least values and again increases. The second section of the graph is reduced as the value of w increases and the repeated increase in the average velocity becomes less noticiable. The second type of graphs holds for w>0, on which an insignificant decrease in the average vertical velocity is observed as φ_0 increases and the greater the value of w, the more significant the decrease of $\dot{y}_{sr}$. The average velocity increases proportionally to φ_0 on the graphs up to the maximum value and after that $\dot{y}_{sr}$ decreases and the value of the last section and the value by which the average velocity decreases both decrease as w increases. The graph for w=0 is a transition between the graphs for w<0 and w>0.

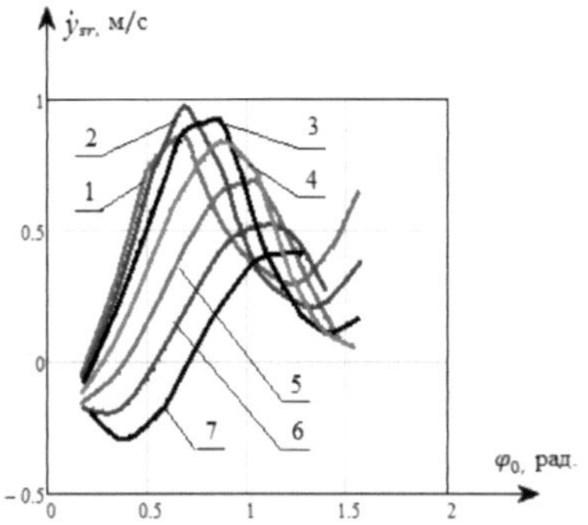

Fig. 6. Graphs of $\dot{y}_{sr}(\varphi_0)$ for λ=1, w=0 rad, T=0,03 s: 1 - w=-15^0, 2 - w=-10^0, 3 - w=-5^0, 4 - w=0^0, 5 - w=5^0, 6 - w=10^0, 7 - w=15^0.

6 Conclusion

In this article we presented the results of the study of the vertical flight of a five-link robot-ornithopter with flapping folding wings whose main distinctive feature is that the motion of its wings imitates the flaps of a seagull. A mathematical model of the flight of the device in the vertical plane was developed and presented in matrix form. We discussed in detail the interaction of the links of the device with the air enviroment, determined the positions and velocities of the points of application of reduced aerodynamic forces to each link of the robot.

Numerical modelling of the robot's flight was conducted based on the mathematical model and three modes of flight: descent, hovering and ascet were established based on the time graphs of of the vertical coordinate of the body's center of mass. Hovering is observed for each value of φ_0 only at a given value of ω, while ascent and descent have a wide range of variation of the two stated parameters.

The graphs of the average values of the body's vertical velocities against the amplitudes of the links' oscillations were constructed and the influence of the osillation parameters: amplitude ratios of the oscillations of links 1 and 2 (4 and 5) and the displacement, w of the equillibrium position of the oscillations from zero on the alternation sequence of the flight modes, maximum and minimum values of $\dot{y}_{sr}$.

The study was conducted with the financial support of RFBR under scientific project № 16-08-00787.

References

1. J.D. DeLaurier, *Canadian aeronautics and space journal*, **40**, 1, 10–18 (1994)
2. A.N. Brooks, *AIAA*, 85, 1446 (1985)
3. J.D. DeLaurier, *The aeronautical journal of the royal aeronautical society*, 153–161 (1993)
4. E. Craparo, B.A Ingram, *41st aerospace sciences meeting and exhibit*, p. 1-9 (2003)
5. S.F. Jatsun, L.Yu. Vorochaeva, S.V, *Vibroengineering PROCEDIA*, **8**, 74-79 (2016)
6. Z.J. Jackowski, Diss. Massachusetts Institute of Technology (2009)
7. J. Maglasang, K. Isogai, N. Goto, M. Yamasaki, *Memoirs of the Faculty of Engineering*, **66**, 1, 71-82 (2006)

Optimal Stabilization of A Quadrotor UAV by a Constrained Fuzzy Control and PSO

Hamid Boubertakh[1, *]

University MSBY of Jijel, LAJ-Laboratoire d'Automatique de Jijel, B.P. 98 Ouled Aïssa 180000 Jijel, Alegria

Abstract. This work aims to design an optimal fuzzy PD (FPD) control for the attitude and altitude stabilization of a quadrotor. The control design is done by mean of the particle swarm optimization (PSO) under the constraints of the controller interpretability and the saturation of the actuators. Concretely, a decentralized control structure is adopted where four FPD controllers are used to stabilize the quadrotor angles (roll, pitch and yaw) and height. A PSO-based algorithm is used to simultaneously tune the four constrained controllers regarding a cost function quantifying the whole system performances. The simulation results are presented to show the efficiency of the proposed approach.

1 Introduction

In the recent last years, the quadrotor unmanned aerial vehicles (UAVs) are become very familiar. This Famousness is due to their success trough many applications in various domains.
The raisons behind this attraction by quadrotors return largely to their simple and useful configuration with six DOF. The movement of the system is controlled via the relative speeds of four rotors.

The quadrotor stabilization is primordial for a safe and reliable tasks achievement. Therefore, many control techniques have been applied to this end [1–6]. Among them, the classical PID control has been largely used despite its poor performances when it is badly tuned and/or used with nonlinear systems.
Fuzzy PID control [7] is presented in literature as an alternative to address the difficulties with classical PID control. It combines the advantages these controllers namely the simplicity, the interpretability of the control structure and more they can deal with non linear systems.

Last years, the population-based optimization methods have been gained a special interest to solve the fuzzy controllers design problem; Genetic Algorithm (GA) [8], Particle swarm optimization(PSO) [9], Ant Colony Optimization (ACO) [10], Bees Algorithm [11], and Gravitational Search Algorithm (GSA) [12]. Among these methods, PSO is distinguished by the simplicity and the efficiency of the algorithm trough a continuous search space.

In this paper, one suggests to apply the PSO for the fuzzy control optimal design for attitude and altitude stabilization of a quadrotor, under the system inputs saturation physical constraint and the maintain of the logical interpretability in the optimal controllers' structures.

The remainder of this paper is organized as follows. Section II gives a general description and presents a dynamical model of a quadrotor. Section III gives an overview the PSO and presents the FPD controller design method using PSO. In section IV, first the FPD decentralized control structure for the stabilization of the quadrotor is presented then PSO design method is used and simulation results are presented. Section V concludes the paper.

2 The quadrotor model

The quadrotor helicopter is a small UAV with four propellers actuated by DC motors mounted on the end of two perpendicular arms. A basic diagram is shown in Fig. 1. Each rotors pair of the same arm rotates in the same direction; one pair rotates clockwise, while the other rotates counter clockwise. The quadrotor moves by adjusting the angular velocity of each rotor.

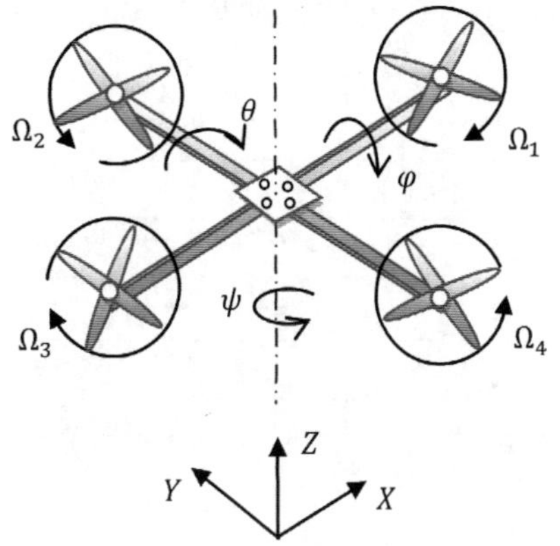

Fig. 1. The quadrotor configuration

* Corresponding author: boubert_hamid@yahoo.com

The quadrotor dynamical model is given by [1]:

$$\begin{aligned}
\dot{x}_1 &= x_2 \\
\dot{x}_2 &= x_4 x_6 a_1 - x_4 a_2 \Omega_r + b_1 u_2 \\
\dot{x}_3 &= x_4 \\
\dot{x}_4 &= x_2 x_6 a_3 + x_2 a_4 \Omega_r + b_2 u_3 \\
\dot{x}_5 &= x_6 \\
\dot{x}_6 &= x_2 x_4 a_5 + b_3 u_4 \\
\dot{x}_7 &= x_8 \\
\dot{x}_8 &= sin(x_3) cos(x_1) * b_4 u_1 \\
\dot{x}_9 &= x_{10} \\
\dot{x}_{10} &= -sin(x_1) b_4 u_1 \\
\dot{x}_{11} &= x_{12} \\
\dot{x}_{12} &= -g + cos(x_3) cos(x_1) b_4 u_1
\end{aligned}$$

Where

- $x_1 = \varphi$ and $x_2 = \dot{\varphi}$ are respectively the roll angle and corresponding angular velocity;
- $x_3 = \theta$ and $x_4 = \dot{\theta}$ are respectively the pitch angle and corresponding angular velocity;
- $x_5 = \psi$ and $x_6 = \dot{\psi}$ are respectively the yaw angle and corresponding angular velocity;
- $x_7 = X$, $x_9 = Y$, and $x_{11} = Z$ are the Cartesian position coordinates.
- $x_8 = \dot{X}$, $x_{10} = \dot{Y}$, and $x_{12} = \dot{Z}$ are translation velocities.
- $\begin{bmatrix} u_1 \\ u_2 \\ u_3 \\ u_4 \end{bmatrix} = \begin{bmatrix} 0 & b & 0 & -b \\ -b & 0 & b & 0 \\ d & -d & d & -d \\ b & b & b & b \end{bmatrix} \begin{bmatrix} \Omega_1{}^2 \\ \Omega_2{}^2 \\ \Omega_3{}^2 \\ \Omega_4{}^2 \end{bmatrix}$
- $\Omega_r = -\Omega_1 + \Omega_2 - \Omega_3 + \Omega_4$ is the sum of rotor angular velocities.
- $a_1 = \frac{Iy-Iz}{Ix}; a_2 = \frac{Jr}{Ix}; a_3 = \frac{Iz-Ix}{Iy}; a_4 = \frac{Jr}{Iy};$
- $a_5 = \frac{Ix-Iy}{Iz}; b_1 = \frac{l}{Ix}; b_2 = \frac{l}{Iy}; b_3 = \frac{1}{Iz}; b_4 = \frac{1}{m}.$

The values of the parameters in the state space model are given by Table1.

Table 1. The quadrotor model parameters

Parameter	Value
l	$0.232\ m$
d	$7.5 \times 10^{-7}\ N.m.s^2$
b	$3.13 \times 10^{-5} N.s^2$
m	$0.52\ Kg$
g	$9.806\ m/s^2$
I_x	$6.228 \times 10^{-3} Kg.m^2$
I_y	$6.228 \times 10^{-3} Kg.m^2$
I_z	$1.121 \times 10^{-2} Kg.m^2$
J_R	$6.01 \times 10^{-5} Kg.m^2$

3 The fuzzy PD control

The fuzzy proportional derivative (FPD) control law is given by

$$u_{FPD}(t) = f(e(t), \Delta e(t)) \quad (1)$$

With ; $e(t) = y_r(t) - y(t)$ and $\Delta e(t) = \frac{de(t)}{dt}$.
Where f is the function of the fuzzy inference system, $u_{PFD}$is the control signal, y_r is the reference signal and y is the system's controlled output.
According to the typical structural design of FPD controllers in [13], one uses FPD controllers with 3 membership functions for input variables and 5 singletons for the output variable as shown on Fig. 2.
The rule base of the controllers is donated by Table 2.
The crisp control action is determined by the average sum (2).

$$u_{FPD}(t) = \frac{\sum_{i=1}^{9} V_i C_i}{\sum_{i=1}^{9} V_i}. \quad (2)$$

Where C_i is the conclusion of rule i and V_i is its truth value calculated by the algebraic product method given by: $V_i = \mu_{e_i}(e(t)).\mu_{\Delta e_i}(\Delta e)$. $\mu_{y_i}(y)$ is the membership grade of the input variable y evaluated in rule i by the corresponding membership function.

Table 2. FPD controller rule base

		Δe	
e	N	Z	P
N	C_1=NB	C_2=NS	C_3=ZR
Z	C_4=NS	C_5=ZR	C_6=PS
P	C_7=ZR	C_8=PS	C_9=PB

The labels NB, NS, ZR, PS, and PB refer respectively to the linguistic terms; Negative Big, Negative Small, around Zero, Positive Small and Positive Big.
Note here that, if one considers the typical structure design of the FPD controller with

- The input membership functions and the output singletons are evenly distributed on their corresponding universes of discourse, i.e., $E_P = E_N, D_P = D_N$ and $U_{NB} = -U_{PB} = 2U_{NS} = = -2U_{PS} = U$.
- The universes of discourse are large enough such that all possible inputs are always within their limits.

Then, the crisp control signal calculated by (2) is given by [13]:

$$u_{FPD}(t) = \frac{U}{2E} e(t) + \frac{U}{2D} \frac{de(t)}{dt}. \quad (3)$$

Equation (6) shows clearly that the typical FPD controllers are identical to the classical PD controllers with the following proportional and integral gains: $K_P = \frac{U}{2E}$ and $K_D = \frac{U}{2D}$.

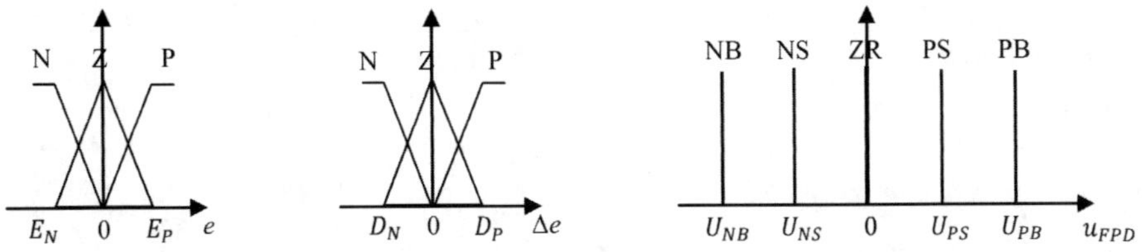

Fig 2. Membership functions of the FPD controllers

A nonlinear FPD controller can be reached by moving the input membership functions and/or the output singletons from their modal positions. That is, the FPD controllers combine the advantages of classical PD controllers and fuzzy controllers; the simplicity, the structure interpretability, and the nonlinearity.

4 Quadrotor optimal stabilization using PSO design-based fuzzy control

4.1 Particle swarm optimization

Particle swarm optimization (PSO), initially introduce by Kennedy and Eberhart [14], is a heuristic optimization method inspired by the movement of individuals within a swarm. PSO technique conducts search in parallel using a population of particles, corresponding to individuals. Each particle represents a candidate solution to the problem at hand. In the PSO system, a particle makes use of the best position encountered by itself and that of its neighbors to steer toward an optimal solution. The performance of each particle is measured using a predefined fitness function, which quantifies the performance of the optimization problem.
The mathematical expressions for velocity and position updates are given by

$$v_{ij}^{k+1} = wv_{ij}^{k} + c_1r_1(Pbest_{ij} - x_{ij}^{k}) + c_2r_2(Gbest_j - x_{ij}^{k}) \quad (4)$$

$$x_{ij}^{k+1} = x_{ij}^{k} + v_{ij}^{k+1} \quad (5)$$

Where: $1 \le i \le N,\ 1 \le j \le D, 1 \le k \le K$.
N is the number of particles in the swarm, D is the dimension of particle, and K is the maximum iterations. x_{ij}^{k} is the position of particle i in the dimension j at iteration k, v_{ij}^{k} is the velocity of particle i in the dimension j at iteration k, $pbest_{ij}$ is a personal best of particle i in the dimension j, $gbest_j$ is a global best of all particles in the dimension j. w is the inertia weight factor, c_1 and c_2 are the acceleration constants, r_1 and $r_2$are random numbers in interval $[0, 1]$.

4.2 Application to the stabilization of the quadrotor by fuzzy control

4.2.1. The control structure

The control goal is to stabilize angles and height of the quadrotor. That is, one uses four decentralized FPD controllers as presented by the bloc diagram on the Fig. 3. The controllers $\mathrm{FPD_i}$ (i = 1,2,3) have the same structure of section 3, however, the $\mathrm{FPD_4}$ is modified as:

$$\mathrm{u_4(t) = mg + u_{FPD}(t)}$$

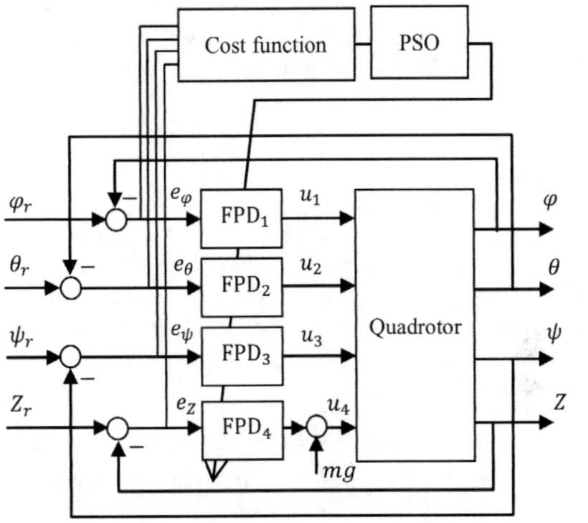

Fig. 3 Decentralized FPD control

4.2.2. The parameters' design vector

The design parameters' vector is constructed by the positions of the input membership functions and the positions of the output singletons of the four FPD controllers. However, one takes symmetrical fuzzy controllers with respect to zero, i.e.,$E_{Pi} = -E_{Ni} = E_i, D_P = -D_N = D_i, U_{PSi} = -U_{NBi} = U_{1i}$ and $U_{PBi} = -U_{NBi} = U_{2i}$ where the index i refer to controller $\mathrm{FPD_i}$. In addition, because of the quadrotor body symmetry, one takes identical controllers for the roll and pitch angles (i.e. $\mathrm{FPD_1 = FPD_2}$).
That is, the parameters' vector is then

$$P = [E_1\ D_1\ U_{11}\ U_{12}\ E_3\ D_3\ U_{31}\ U_{32}\ E_4\ D_4\ U_{41}\ U_{42}]^T$$

4.2.3. The cost function

The effectiveness of the FPD control scheme is evaluated at the end of the whole control system response to desired inputs by a unique cost function evaluating simultaneously the four system responses. Therefore, in order to overcame the non homogeneity of the outputs, in the cost function instead of using the errors signals directly, one uses the relative errors defined as : $e_1(t) = \frac{e_\varphi(t)}{e_\varphi(t_0)}$, $e_2(t) = \frac{e_\theta(t)}{e_\theta(t_0)}$, $e_3(t) = \frac{e_\psi(t)}{e_\psi(t_0)}$ and $e_4(t) = \frac{e_Z(t)}{e_Z(t_0)}$. Without loss of generality, one considers in this work the integral sum of relative errors (ISRE) cost function given by

$$ISRE = \int_{t_0}^{t_f} (e_1^2(t) + e_2(t)^2 + e_3^2(t) + e_4^2(t))\, dt$$

4.2.4. The optimization problem constraints

In this work one consider two types of constraints.

- *The control structure interpretability*

One has to preserve the semantic interpretability of the fuzzy rule bases of the controllers such that all parameters values correspond concretely to their linguistic terms (e.g.: a minus signed value correspond to the term 'Negative', a positive signed value correspond to the term 'Positive ', a 'Small' value is less than a 'Medium' value which is itself less than a 'Big' value,...).

- *The system input constraints*

In practice, the system inputs are subject to physical constraints often caused by the actuators limitations. That can degrade the closed system performance and may leads to the instability, if one does not take care of the problem in the control design. In this work, one considers the actuators' amplitudes saturation such as a control inputs u_i do not come over certain bounds, i.e. $u_{min_i} \le u_i \le u_{max_i}$.

$$u_i = \begin{cases} u_{max_i} & if \quad u_i > u_{max_i} \\ u_{min_i} & if \quad u_i < u_{min_i} \\ u_i & otherwise \end{cases}$$

In this work one takes:

$$-1 \le u_{FPDi} \le +1 \ , (i = 1, \ldots, 3)$$

$$0 \le u_{FPD4} \le +15$$

4.2.5. The algorithm

In the algorithm of the FPD control design using PSO, a particle is one value of the FPD controllers' parameters vector. A particle swarm is a group of parameters vector values which are manually or randomly initialized in search space defined by limits of the components of the tuning parameters vector. So, a particle position x is the controllers' parameter vector value, $Pbest$ is the best parameters values vector regarding the cost function, $Gbest$ is the global best parameters values vector of all particles. The algorithm steps are summarized in table 3.

4.2.6. Simulations

In the simulations, one takes:

- The initial states: $\varphi(0) = 0.5\, rad$, $\theta(0) = -0.5\, rad$, $\psi(0) = 0.5\, rad$, and $Z(0) = 0\, m$.
- The desired angles and height :$\varphi_d = 0\, rad$, $\theta_d = 0\, rad$,$\psi_d = 0\, rad$, and $Z_d = 10\, m$,
- The cost function spans from $t_0 = 0$ sto $t_f = 20\, s$.
- The PSO algorithm parameters: $\omega = 0.9$, $c_1 = 2$, $c_2 = 2$, $N = 10$, $K = 100$.
- The initial parameters values are chosen randomly between the following bounds: $0 \le E_i \le 10$, $0 \le D_i \le 10$, $0 \le U_{i1} \le 15$, $15 \le U_{i2} \le 30$. (i=1,...4).

The Fig. 4 shows the evolution of the cost function with iterations. Table 4 gives the optimal parameters' values. Fig. 5 show the angles (roll, pitch and yaw), Fig. 6 show the control signals u_1, u_2 and u_3. Fig 7 show the height and Fig. 8 show the control signal u_4.

Table 3. The constrained PSO algorithm

```
Begin
k=0
Initialize all particles positions and speeds
  Repeat form k=1:K
        For each particle
          – Compute the controllers actions u_FPDi
            (i=1,...4)
          – Apply the control actions to de system.
          – Evaluate the cost function of the particle.
          If the constraints are not all satisfied
             – Attribute the worst value to the
               corresponding cost function.
          End
        – Evaluate Pbest
        End
        – Evaluate Gbest
        For each particle
          – Update the particles' speeds and
            positions by (4) and (5) respectively.
        End
        k←k+1
  End
The optimal parameters' vector is Gbest.
End
```

Table 4. The optimal paramers' values

Parameter	Optimal value
$E_1 = E_2$	2.5077
$D_1 = D_2$	5.3927
$U_{11} = U_{21}$	4.5098
$U_{12} = U_{22}$	41.9965
E_3	2.6048
D_3	21.9211
U_{31}	3.5153
U_{32}	15.531
E_4	3.6861
D_4	5.8867
U_{41}	8.5665
U_{42}	19.0788

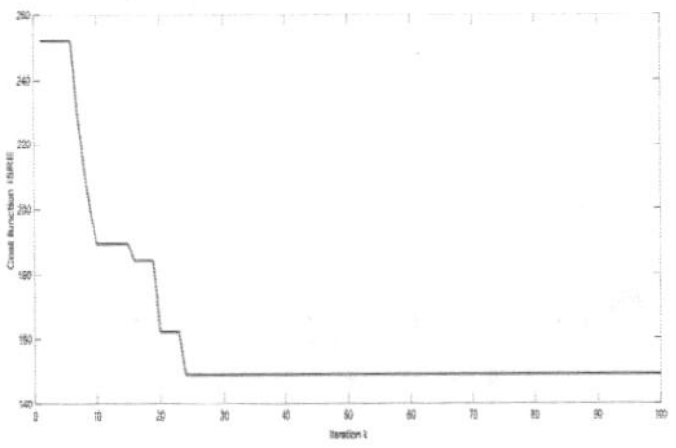

Fig 4. Cost function

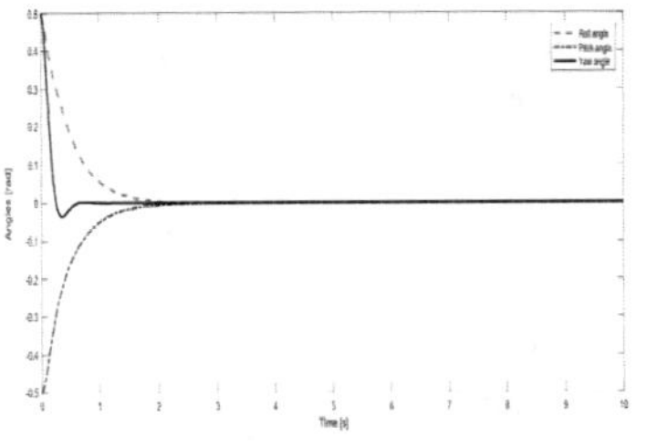

Fig. 5 Angles φ, θ and ψ

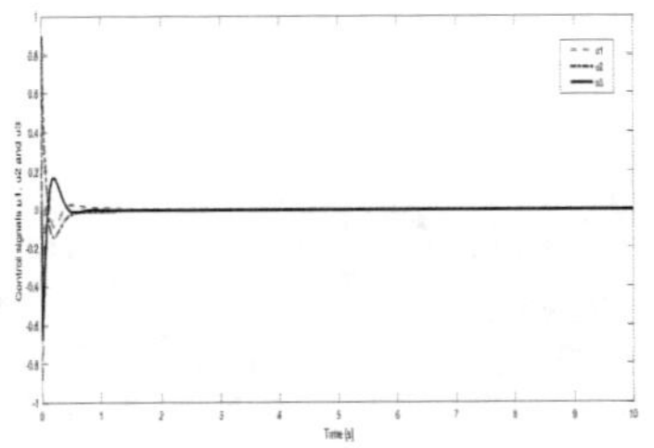

Fig. 6 Control signals u_1, u_2 and u_3

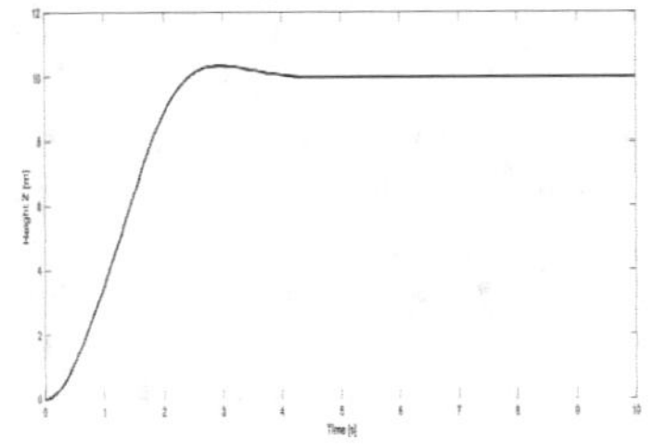

Fig 7. Height Z

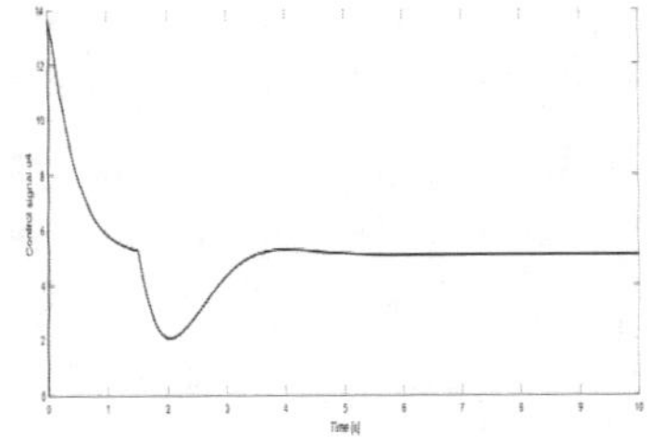

Fig 8. Control signal u_4

5 Conclusion

In this paper, one has proposed a PSO-based method to FPD control design for the stabilization of a quadrotor under the actuators saturation and interpretability of the FPD control. Four FPD controllers are used in a decentralized control structure to stabilize the quadrotor attitude and height. The constrained PSO method is used to tune the FPD controllers' parameters simultaneously in order to minimize an integral square error cost function which quantity the whole system performance. The simulation results show the efficiency of the proposed method.

References

1. S. Bouabdallah, *Design and control of quadrotors with application to autonomous flying* (Ph.D. diss. École Polytechnique Fédérale de Lausanne, 2007).
2. T. Madani, A. Benallegue, Backstepping control for a quadrotor helicopter, *IEEE International Conference on Intelligent Robots & Systems*, 3255-3260 (2006).
3. A. Tayebi and S. McGilvray, *IEEE Trans. Syst. Tech.*, **14**, 552-570 (2006)
4. P. Castillo, A. Dzul, R. Lozano, *IEEE Trans. Syst., Man, Cybern. B*, **32**, 239-248 (2002).
5. K. U. Lee, H. S. Kim and Y.-H. Choi, Hovering control for a quadrotor, *12th Conference on Control, Automation and Systems*, 162-167 (2012)
6. H. Bolandi, M. Rezaei, M. Smailzadeh, Intelligent Control and Automation, 4, 335-342 (2013).
7. R.R. Yager, D.P. Filev, *Essential of fuzzy modeling and control* (John Wily & Sons Inc, 1995)
8. J.M. Adams, K.S. Rattan, A genetic multi-stage fuzzy PID controller with a fuzzy switch, *Proc. IEEE Int. Conf. on Systems, Man, Cybernetics, Tucson*, AZ, USA, **4**, 2239-2244 (2001)
9. H. Boubertakh, S. Labiod and M. Tadjine, PSO to design decentralized fuzzy PI controllers: Application for a Helicopter, *19th Mediterranean Conference of Control and Automation (MED'12)*, IEEE, pp. 633-637 (2012)
10. H. Boubertakh, M. Tadjine, P.-Y. Glorennec, S. Labiod, Tuning fuzzy PID controllers using ant colony optimization, *17th Mediterranean Conference on Control and Automation (MED'09), IEEE*, 13–18 (2009)
11. C. Craveo, F. Valdez, O. Castillo, *Appl. Soft Comp.*, **43**, 131-142 (2016)
12. C.-T. Chao, M.-T. Liu, J.-S. Chiou, Y.-J. Aided *Engineering and Software*, **33**, 1659-1667 (2016)
13. H. Boubertakh, M. Tadjine, P.Y. Glorennec, S. Labiod, *ISA Trans.*, **49**, 543-551 (2010)
14. J. Kennedy, R. Eberhart, Particle swarm optimization, *IEEE International Conference on Neural Networks*, pp. 1942-1948 (1995)

DOI: 10.1051/matecconf/20179903002

Algorithms of the Objects' Identification According to Hyperspectral Shooting of the „Resource-P" Spacecraft

Liliya Demidova[1,2] and *Sergey* Trukhanov[3,*]

[1]Ryazan State Radio Engineering University, 390005 Ryazan, Russia
[2]Moscow Technological Institute, 119334 Moscow, Russia
[3]Branch of JSC „SRC „Progress" – Special Design Bureau „Spectr", 390005 Ryazan, Russia

Abstract. The identification algorithms, which carry out the objects' identification of the Earth's surface by means of their hyperspectral features' analysis, received on the base of the processed space images from the „Resource-P" spacecrafts with application of the similarity measures, have been considered. The identification algorithms on the base of the Euclidean distance similarity measure, the angular similarity measure and the fuzzy similarity measure have been applied. The use expediency of the fuzzy linear regression in the algorithm of objects' hyperspectral features' identification has been shown. The results of the hyperspectral information processing with using of the offered algorithms have been presented.

1 Introduction

The problem of processing and analysis of hyperspectral information formed on the base of hyperspectral images of the Earth's surface presented by a big set of pictures of the same scene in the narrow adjoining spectral ranges received from a board of the spacecraft is one of the actual tasks solved by systems of Earth remote sensing. So, in particular, the russian „Resource-P" spacecrafts No. 1 – 3 with the hyperspectral equipment (HSE) on the board started respectively in 2013 – 2016, give out sets of pictures in 130 narrow adjoining ranges of visible area of spectrum which form a hyperspectral image (HSI) [1]. When processing GSI much attention is paid to the questions of the objects' identification of the Earth's surface, using their hyperspectral features (HSF).

Object's HSF in a graphic form can be presented as display of interrelation between the wavelength and values as spectral brightness coefficient (SBC), and spectral reflection coefficient (SRC) of the analyzed object. However at the solution of a problem of object's identification on its HSF use of SRC dependence on wavelength is more preferable as this brightness characteristic doesn't depend on shooting conditions in such degree as SBC. Besides, unlike SBC, for obtaining SRC values the standard reflecting surface in sight of the analyzed object isn't necessary.

Object's identification on its HSF can be carried out by means of comparison with application of some similarity measures of the analyzed object's HSF with some standard HSFs united in special spectral libraries [2–4]. Thus the analyzed and standard HSFs must have identical units of measure, spectral range and permission of data [5].

The analysis of works [6–10] testifies that for the problem solution of objects' identification of the Earth's surface on their HSFs the following approaches are most often used: algorithms on the base of similarity measures; method of a spectral corner; artificial neural networks; algorithm of k-averages; method of maximum likelihood.

The materials of experimental studies, received with application of the above approaches testify that we haven't algorithms which provide the demanded quality of objects' identification on their HSF in the prevailing majority of cases. Thus, the development problem of identification's algorithms of objects' HSF of the Earth's surface with application of various reasonably chosen similarity measures, and also of the aggregation algorithm of identification's private results of objects' HSF received when using several identification's private algorithms of objects' HSF is actual [11–16].

For the solution of identification's problem of objects on its HSFs it is offered to use at the same time four algorithms of HSF identification based on Euclidean distance similarity measure, angular similarity measure and two fuzzy similarity measures for the purpose of the subsequent aggregation of identification's private results of objects' HSF.

The choice of Euclidean distance similarity measure can be proved by high efficiency of application of the given measure at the solution of various applied problems of identification (classification), including problems of identification when processing space images. The choice of angular similarity measure assuming realization of spectral corner method (Spectral Angle Mapper – SAM), can be proved by that this measure provides good identification's results of objects' HSF having similar values of points' brightness of the

* Corresponding author: serge_tsv@mail.ru

image in all spectral ranges even when on initial images effects of a flare are observed. This fact is explained by that the spectral corner method doesn't consider brightness values of image points. Application expediency of fuzzy similarity measures [5, 6] can be proved by that the analyzed HSF of object can belong at the same time to different classes of objects while algorithms of «rigid» identification which are algorithm on the base of Euclidean distance similarity measure and algorithm on the base of angular similarity measure, will unambiguously carry the analyzed object's HSF to any one unambiguously certain class that can be not always true. The algorithms of „soft" identification based on application of fuzzy similarity measures will allow to solve flexibly a problem of object's identification using HSFs.

2 Identification of object's hyperspectral characteristics on the base of Euclidean distance similarity measure

Identification of object's HSF on the base of Euclidean distance similarity measure realizes identification of HSF with the use of the Euclidean metrics for calculation of distance between two points in the J-dimensional space [15, 16]:

$$E = \left(\sum_{j=1}^{J} (y_j^A - y_j^S)^2 \right)^{0,5}, \quad (1)$$

where y_j^A is the SRC value of the analyzed HSF for the j-th hyperspectrometer channel; y_j^S is the SRC value of the standard HSF for the j-th hyperspectrometer channel; $j = \overline{1,J}$; J is the hyperspectrometer channel's number, equal to points' number (measurements) for HSF (for example, $J = 96$).

Wavelength λ_j for j-th hyperspectrometer channel ($j = \overline{1,J}$) is considered as well known value. The SRC values of the analyzed and standard HSFs are put in compliance to wavelengths.

Algorithm of HSF identification on the base of Euclidean distance similarity measure assumes: calculation of Euclidean distance similarity measures (1) for the analyzed HSF and standard HSFs, stored in the database; ordering on increase of the calculated values of Euclidean distance similarity measures; choice as required that standard HSF for which value of Euclidean distance similarity measures (1) is minimum.

3 Identification of object's hyperspectral characteristics on the base of angular similarity measure

Identification of the object's HSF on the base of angular similarity measure assumes use of spectral corner method (Spectral Angle Mapper – SAM) realizing an assessment of similarity of the analyzed and standard HSF, considered as vectors which dimension is equal to number of hyperspectrometer channels, by means of corner's calculation between them [5]:

$$\alpha = arccos \frac{G \cdot G'}{|G| \cdot |G'|} = arccos \frac{\sum_{j=1}^{J} g_j \cdot g'_j}{\left(\sum_{j=1}^{J} g_j^2 \right)^{0,5} \cdot \left(\sum_{j=1}^{J} g_j'^2 \right)^{0,5}}, \quad (2)$$

where G and G' are the spectrums of the analyzed and standard HSFs respectively; g_j and g_j' are the SRC values of the analyzed and standard HSFs respectively for wavelength's value λ_j ($j = \overline{1,J}$); J is the hyperspectrometer channel's number.

Algorithm of HSF identification on the base of angular distance similarity measure assumes: calculation of angular similarity measures (2) for the analyzed HSF and standard HSFs, stored in the database; ordering on increase of the calculated values of angular distance similarity measures; choice as required that standard HSF for which value of angular distance similarity measures (2) is minimum.

Often because of errors of HSE, resulting in inaccuracy of information on the about the analyzed object's HSF and also because of almost total absence of the „pure" analyzed HSF and existence in most cases of the analyzed HSF representing mix from several basic classes [5], it is expedient to consider some standard HSFs having the smallest values of Euclidean distance similarity measure (1) and angular similarity measure (2) as the potentially required. However, despite use at identification of a large number of the standard HSFs, results of identification with application of Euclidean distance similarity measure or angular similarity measure can be unsatisfactory.

In this regard it is expedient to carry out confirmation of identification's result, which, in particular, can be received by means of application of other identification's algorithms of object's HSF with the subsequent aggregation of identification's private results. As show experimental studies, the good aggregating identification's result provides sharing of HSF identification's algorithms on the base of Euclidean distance similarity measure, on the base of angular similarity measure and on the base of fuzzy similarity measures.

4 Identification of object's hyperspectral characteristics on the base of fuzzy similarity measures

Identification of object's HSF can be executed on the base of classical linear regression's equation (CLR) by means of the problem's solution of equation parameters' selection (k and b) for the analyzed HSF and standard HSF (for example, according to method of least squares):

$$y = k \cdot x + b \quad (3)$$

with the subsequent calculation of root-mean-square deviation (RMSD) for the analyzed HSF and standard HSF, which can be used as the characteristics of uniqueness at HSF identification.

If for object's HSF identification we use only one characteristic of uniqueness – such as RMSD, in some cases objects' HSF, having approximately equal values of RMSD, can be carried to one class in spite of the fact that the form of curve for objects' HSF will be various. In this regard for identification of object's HSF it is offered to use additional characteristics of uniqueness which can be received on the base of fuzzy linear regression (FLR) equation. These characteristics can be used for calculation of fuzzy similarity measures.

In [6] it is shown that is most expedient to use FLR equation with asymmetric fuzzy parameters as it (unlike FLR equation with symmetric fuzzy parameters) provides calculation of the uniqueness characteristic's value – RMSD, which is equal or close to the value received by means of CLR equation:

$$Y(x) = A_1 \cdot x + A_0 , \qquad (4)$$

where $A_1 = (a_1, c_1, d_1)$ and $A_0 = (a_0, c_0, d_0)$ – triangular fuzzy numbers (TFN), corresponding to parameters k и b of CLR equation (3), represented by means of the triangular membership functions, and considered as the asymmetric fuzzy parameters of FLR equation (4).

When algorithms' developing of object's HSF identification on the base of fuzzy measures various fuzzy similarity measures [6] can be used. Thus identification algorithms will have identical stages of realization.

At the first stage of algorithm parameters (namely TFN) the FLR equation of the analyzed HSF are defined. Therefore the problem of square programming (PSP) formulated as follows [7] is solved:

$$F_{ЗКП} = k_1 \cdot \sum_{j=1}^{J} (y_j - \sum_{i=0}^{n} a_i \cdot x_{ji})^2 + = k_2 \cdot (1-\alpha) \cdot \sum_{j=1}^{J} \sum_{i=0}^{n} (c_i + d_i) \cdot x_{ji} + \xi \cdot \sum_{i=0}^{n} (c_i^2 + d_i^2) \to \min_{a,c,d} \qquad (5)$$

and

$$\sum_{i=0}^{n} a_i \cdot x_{ji} + (1-\alpha) \cdot \sum_{i=0}^{n} d_i \cdot x_{ji} \geq y_j ; \qquad (6)$$

$$\sum_{i=0}^{n} a_i \cdot x_{ji} - (1-\alpha) \cdot \sum_{i=0}^{n} c_i \cdot x_{ji} \leq y_j ; \qquad (7)$$

$$c_i \geq 0; d_i \geq 0 \ (j = \overline{1,J};\ i = \overline{0,n};\ n = 1) , \qquad (8)$$

where j is the ordinal number of hyperspectrometer channel ($j = \overline{1,J}$); J is the hyperspectrometer channel's number; i is the ordinal number of TFN (if $n = 1$ then $i = \overline{0,1}$); $x_{j0} = 1$ ($j = \overline{1,J}$) – parameter b of CLR equation (2); x_{j1} – wavelength value for the j-th hyperspectrometer channel; y_j is the SRC value for wavelength x_{j1} ($x_{j1} = \lambda_j$); α is a value of TFN level, characterizing FLR corridor's width ($\alpha \in [0,1]$); k_1, k_2 are the weight coefficients characterizing a contribution of the first and second items in the objective function (5); ξ is a small positive number such that $k_1, k_2 >> \xi$ (the third item is entered into criterion function in order that it had a square type and that at the search of TFN values could formulate the PSP [17]).

At the solution of PSP (5)–(8) it is supposed that $k_1 = k_2 = 1$ (in [7] it is shown that the choice of parameters' values k_1, k_2 has no strong impact on the problem solution); $\xi = 0{,}001$. Let $a_0 = z_1$; $c_0 = z_2$; $d_0 = z_3$; $a_1 = z_4$; $c_1 = z_5$; $d_1 = z_6$; $x_{j1} = \lambda_j$; $y_j = g_j$.

Then PSP (5) – (8) can be written as:

$$F_{ЗКП} = \sum_{j=1}^{J} (g_j - z_1 - z_4 \cdot \lambda_j)^2 + \sum_{j=1}^{J} (z_2 + z_5 \cdot \lambda_j + z_6 \cdot \lambda_j) + + 0{,}001 \cdot (z_2^2 + z_3^2 + z_5^2 + z_6^2) \to \min_{z_1, \ldots z_6} \qquad (9)$$

and

$$\begin{cases} z_1 + z_3 + z_4 \cdot \lambda_j + z_6 \cdot \lambda_j \geq g_j; \\ z_1 - z_2 + z_4 \cdot \lambda_j - z_5 \cdot \lambda_j \leq g_j; \\ z_2 \geq 0; \\ z_3 \geq 0; \\ z_5 \geq 0; \\ z_6 \geq 0. \end{cases} \qquad (10)$$

For the solution of PSP in the form of (8) at restrictions (9) the method of Lagrange uncertain multipliers can be used. In view of the fact that the solution of PSP by this method is very labor-consuming problem, use of the existing mathematical packages allowing to solve minimization problems with restrictions is expedient. In particular, can be solved in system of engineering and scientific calculations of MATLAB with application of the built-in function „quadprog" [18].

FLR equation (3) for the analyzed object's HSF is formed on the base of PSP solution (9) – (10). Thus on the base of the calculated TFNs $A_1 = (a_1, c_1, d_1)$ и $A_0 = (a_0, c_0, d_0)$ of FLR equation (4) for values of waves' lengths $x = \lambda_j$ ($j = \overline{1,J}$) characteristic points of the CLR equation $Y_{FLR}^{CLR}(\lambda_j)$ is defined:

$$Y_{FLR}^{CLR}(\lambda_j) = a_0 + a_1 \cdot \lambda_j , \qquad (11)$$

Also the equations of the upper $Y_{FLR}^{UP}(\lambda_j)$ and lower $Y_{FLR}^{LOW}(\lambda_j)$ borders of FLR corridor of the analyzed characteristic are defined:

$$Y_{FLR}^{UP}(\lambda_j) = a_0 + d_0 + (a_1 + d_1) \cdot \lambda_j , \qquad (12)$$

$$Y_{FLR}^{LOW}(\lambda_j) = a_0 - c_0 + (a_1 - c_1) \cdot \lambda_j . \qquad (13)$$

The FLR equation (3) for the standard HSF, and also the CLR equation, the upper $Y_{FLR}^{UP}(\lambda_j)$ and lower $Y_{FLR}^{LOW}(\lambda_j)$ borders of FLR corridor, defined according to (11), (12) and (13), are formed similarly.

Herewith, all calculated values of FLR for the standard HSF are stored in the database.

The FLR corridors of the analyzed HSF and standard HSF are asymmetric (fig. 1) because of TFN's $A_1 = (a_1, c_1, d_1)$ and $A_0 = (a_0, c_0, d_0)$ asymmetry.

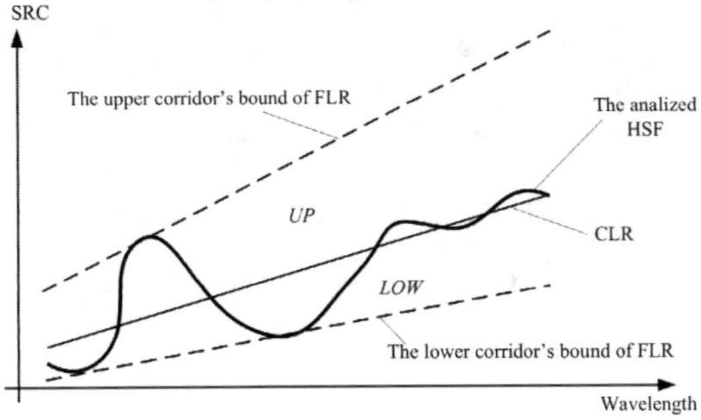

Fig. 1. The HSF representation in FLR corridor.

In this regard the HSF points can be broken into 2 subsets: a subset of the points lying in the upper part of FLR corridor (between the line of upper bound of FLR corridor and the CLR line, determined respectively by the equations (12) and (11)), and a subset of the points lying in the lower part of FLR corridor (between the line of the lower bound of FLR corridor and the CLR line, determined respectively by the equations (13) and (11)).

At the second stage of algorithm for the points of the analyzed HSF and each standard HSF lying in the upper *UP* and lower *LOW* parts of FLR corridors we calculate values of fuzzy similarity measures F^{UP} and F^{LOW}, using one of two fuzzy similarity measures [6]:

$$f_1 = 1 - \frac{\sum_{j=1}^{J} | u_A(\lambda_j, g'_j) - u_S(\lambda_j, g''_j) |}{\sum_{j=1}^{J} | u_A(\lambda_j, g'_j) + u_S(\lambda_j, g''_j) |}, \qquad (14)$$

$$f_2 = \frac{1}{J} \cdot \sum_{j=1}^{J} \frac{min(u_A(\lambda_j, g'_j), u_S(\lambda_j, g''_j))}{max(u_A(\lambda_j, g'_j), u_S(\lambda_j, g''_j))}, \qquad (15)$$

where $u_A(\lambda_j, g'_j)$ is a value of membership function of fuzzy set A of the analyzed HSF to the FLR equation of this HSF for SRC value g'_j, corresponding to wavelength λ_j ($j = \overline{1, J}$); $u_S(\lambda_j, g''_j)$ is a value of membership function of fuzzy set S of the standard HSF to the FLR equation of this HSF for SRC value g''_j, corresponding to wavelength λ_j ($j = \overline{1, J}$); J is the channels' number of the hyperspectrometer equal to points' number in the analyzed (standard) HSF.

These similarity measures were chosen from a set of the fuzzy similarity measures, because they showed the highest quality of identification results for the test data sets described by means of FLR equations.The value of membership function of the some HSF point (λ_j, g_j) to the FLR equation of this HSF can be defined as [7]:

$$u(\lambda_j, g_j) = \begin{cases} 1 - \frac{a_0 + a_1 \cdot \lambda_j - g_j}{c_0 + c_1 \cdot \lambda_j}, \text{if} \\ a_0 + a_1 \cdot \lambda_j - c_0 - c_1 \cdot \lambda_j \le g_j \le a_0 + a_1 \cdot \lambda_j; \\ 1 - \frac{g_j - a_0 - a_1 \cdot \lambda_j}{d_0 + d_1 \cdot \lambda_j}, \text{if} \\ a_0 + a_1 \cdot \lambda_j \le g_j \le a_0 + a_1 \cdot \lambda_j + d_0 + d_1 \cdot \lambda_j; \\ 0, \text{otherwise}; \end{cases} \qquad (16)$$

where $a_0, c_0, d_0, a_1, c_1, d_1$ are the TFN's parameters calculated at the solution of PSP (9) – (10).

At the finishing third stage of algorithm for the analyzed HSF and each standard HSF at first the calculation of the resultant fuzzy similarity measure, defined as minimum of fuzzy similarity measures F^{UP} and F^{LOW}, is carried out:

$$F = min(F^{UP}, F^{LOW}). \qquad (17)$$

Then all standard HSFs are ordered on decrease of fuzzy similarity measures' values. Thus as the required we choose that standard HSF, for which fuzzy similarity measures' value (17) is maximum.

Taking into account the stages described above it is possible to tell that HSF's identification algorithm on the base of fuzzy similarity measure assumes:

- finding of FLR equation for the analyzed HSF;
- extraction from the database of the calculated FLR values for the standard HSFs and calculation of fuzzy similarity measures' values F^{UP} and F^{LOW} according to (14) or (15) for the analyzed and standard HSFs;
- calculation of fuzzy similarity measures' values (17) for the analyzed and standard HSFs and a choice as required that standard HSF for which value of fuzzy similarity measures' value (17) is maximum.

5 The aggregation algorithm of identification's private results of object's HSF

As it was already noted, for improvement of identification's quality of the analyzed object's HSF the aggregation realization of identification's private results (in one way or another) is necessary.

At application of four identification's algorithms to the analyzed HSF and some standard HSFs four values are calculated: value of Euclidean distance similarity measure E (1); value of angular similarity measure α (2); two values of fuzzy similarity measures and, the formulas (14), (15) F_1 and F_2 (17) calculated according to (14), (15) и (17).

During the work with the database containing information on the standard HSFs for the analyzed object's HSF the four identifying sets will be received. Each set is the application's result of one of four HSF identification's algorithms and contains K values (according to number of the standard HSFs in the database).

Each such identifying set can be ordered on decrease (increase) of values of the used similarity measure. As a result some rating estimates will be appropriated to the standard HSFs in the database (ordinal numbers in a rating): than value of Euclidean distance similarity measure and angular similarity measure are higher and values of fuzzy similarity measures are lower, the number in a rating is less.

Let R_E^k is the rating assessment of the k -th standard HSF for the identification's algorithm on the base of Euclidean distance similarity measure E ; R_α^k is the rating assessment of the k -th standard HSF for the identification's algorithm on the base of angular similarity measure α ; $R_{F_1}^k$ is the rating assessment of the k -th standard HSF for the identification's algorithm on the base of fuzzy similarity measure F_1 ; $R_{F_2}^k$ is the rating assessment of the k -th standard HSF for the identification's algorithm on the base of fuzzy similarity measure F_2 ($k=\overline{1,K}$).

The aggregation algorithm of identification's private results of object's HSF assumes:

• the aggregation of identification's private results received with application of Euclidean distance similarity measure E , angular similarity measure α and two fuzzy similarity measures F_1 and F_2 by means of formula:

$$\overline{R}^k = (R_E^k + R_\alpha^k + R_{F_1}^k + R_{F_2}^k)/4; \quad (18)$$

• the ordering of the standard HSFs from the database on increase of average values of the rating assessments $\overline{R}^k$ ($k=\overline{1,K}$).

Let the aggregating rating assessments R_{ord}^k are defined as numbers in a rating for the average assessments $\overline{R}^k$ ($k=\overline{1,K}$). It is obvious that for the further analysis it is expedient to use the insignificant part of the best standard HSFs from the database (from 5 to 10 standard HSFs) having the smallest values (ordering numbers) of the aggregating rating assessments R_{ord}^k .

After the visual comparative analysis of identification's results of the analyzed object's HSF by means of some best standard HSFs from the database received in the above way and presented both in text and in a graphic form, the software operator can make a final decision on compliance to the analyzed object's HSF of some standard HSFs from the database.

6 The experimental studies

The developed software was used for solving the problem of identification of the anthropogenous objects according to hyperspectral shooting from the „Resource-P" spacecraft (No. 1 and No. 2).

The experiment for assessing the reliability of identification solutions with the use of the developed algorithms on the real hyperspectral data in the amount of 15 pictures, each of which contained more than 100 objects was conducted.

Table 1. The identification results.

No	The algorithm	The false omissions,%	The false detection,%
1	The Spectral Angle Mapper algorithm (SAM, the program complex ENVI)	15,4	5,2
2	The aggregation algorithm (the developed software)	4,5	5,5

The data sampling of 402 HSFs in which 181 characteristics belong to the anthropogenous objects has been made.

The identification was carried out by the algorithms from the developed software, and also by the SAM algorithm which is realized in the program complex ENVI. The results of identification given in table 1 show the increase in reliability of the identification decision for 10,9% in comparison with the SAM algorithm from the program complex ENVI [18].

7 Conclusions

The results of experimental studies confirm expediency of further development of the offered approach to the solution of identification's problem of objects' HSF, based on aggregation of identification's private results objects' HSF, received with application of various reasonably chosen identification's algorithms for the purpose of reliability's increase of classification decision. Further researches can be directed on reasonable attraction of new similarity measures for the solution of identification's problem and improvement of aggregation algorithm of identification's private results of objects' HSF.

Use of the offered approach for developing HSF identification's algorithms will allow to solve objects' identification's problem of the Earth's surface on allocated from the processed space images of the „Resource-P" spacecraft HSF with the subsequent accumulation of the standard HSFs in the database that, in turn, will provide creation of actual domestic spectral library of standards HSFs which can be applied when monitoring a condition of agricultural grounds, forests, water resources, an ecological condition of soils, etc. [19, 20].

Also, we plan to use the SVM-classifiers in sovling the problem of the HSF identification [21–23]. Herewith, firstly, it is neccassary to create the reprezantative dataset of the HSFs and, secondly, use it to learn the SVM-classifiers. Then, we can identify the new HSF as the HSF of some known class.

References

1. V. Zelentsov, B. Sokolov et al., Resurs-P" Satellite Hyperspectral Data: Preliminary Evaluation of Information Capacities, *Proceedings of the conference Living Planet Symposium*, **740**, 416 (2016)
2. USGS Spectroscopy Lab. http://speclab.cr.us-gs.gov.
3. Jet Propulsion Laboratory, ASTER Spectral Library. NASA. http://speclab.jpl.nasa.gov.
4. A.M. Baldridge, S.J. Hook, C.I. Grove, G. Rivera, *Remote Sensing of Environment*, **113**, 711–715 (2009)
5. C. Yang, J.H. Everitt, J.M. Bradford, *American Society of Agricultural and Biological Engineers*, 51, 729–737 (2008)
6. D. Van der Weken, M. Nachtegael, E.E. Kerre, An overview of similarity measures for images. *Proceedings of ICASSP 2002 (IEEE Int. Conf. Acoustics, Speech and Signal Processing)*, 3317 – 3320 (2002)
7. Lee Haekwan, Tanaka Hideo,. *Journal of the Operations Research Society of Japan*, **42** (1999).
8. J.D. Adams, M.O. Smith, P.E. Johnson, *Journal of Geophysical Research*, **91**, 8090 – 8112 (1996)
9. K. Michael, Su Griffin at al., *Lincoln laboratory journal*, **15** (2005).
10. E.V. Nikulchev, A.P. Kondratov, *Advanced Studies in Theoretical Physics*, **9**, 795-800 (2015)
11. N.N. Astakhova, L.A. Demidova, E.V. Nikulchev, *Applied mathematical sciences*, **9**, 4813–4830 (2015)
12. L.A. Demidova, N.I. Nesterov, Fuzzy and possibilistic segmentation of earth surface images by means of intelligent information technologies. Proceedings *2014 International Conference on Computer Technologies in Physical and Engineering Applications, ICCTPEA 2014*, pp. 35 – 36.
13. L.A. Demidova, *Automation and Remote Control*, **74**, 313–320 (2013)
14. L.A. Demidova, S.V. Truhanov, *Cloud of Science*, **2**, 40–60 (2015)
15. L.A. Demidova, S.V. Truhanov, *Contemporary Engineering Sciences*, **8**(20), 885–900 (2015)
16. J. Dattorro, *Convex optimization & euclidean distance geometry* (Meboo publishing USA, 2005)
17. D.M. Himmelblau, Applied nonlinear programming. (The Univ of Texas, McGraw-Hill Book Company, 1972).
18. S. Attaway, *Matlab A practical introduction to programming and problem solving* (Butterworth Heinemann, 2009).
19. O. Grigorieva, O. Brovkina el al., A complex model for forest ecosystem state assessment based on remote sensing data: case study in Baikalsky Nature Reserve, *The 4th International Workshop on Simulation for Energy, Sustainable Development and Environment, 26-28 September 2016/ Cyprus*, pp. 14-19.
20. V.F. Mochalov, A.V. Markov el al., *Advances in Intelligent Systems and Computing*, **466**, 497-506 (2016)
21. L.A. Demidova, Yu.S. Sokolova, Training set forming for SVM algorithm with use of the fuzzy clustering algorithms ensemble on base of cluster tags vectors similarity matrices, *2015 International Conference "Stability and Control Processes" in Memory of V.I. Zubov (SCP)*, 571-572 (2015)
22. L.A. Demidova, Yu.S. Sokolova, E.V. Nikulchev, *International Journal of Advanced Computer Science and Applications*, **7**(5), 294-312 (2016)
23. L.A. Demidova, Yu.S. Sokolova, E.V. Nikulchev, *International Journal of Advanced Computer Science and Applications*, **7**(2) 16-24 (2016)

Satellite Image Georeferencing on the Base of Coastlines' Fuzzy Masks

Liliya Demidova [1, 2] and *Victor* Ushenkin [1,*]

[1]Ryazan State Radio Engineering University, 390005, Ryazan, Russia
[2]Moscow Technological Institute, 119334, Moscow, Russia

Abstract. The article discusses the possibility of using fuzzy masks in the problem of image georeferencing according to electronic maps. A new approach to membership assessment of the image fragment in fuzzy mask has been suggested. Fuzzy masks have been compared with binary masks by the accuracy of coregistration of image fragments and coastlines extracted from electronic maps. It's shown that fuzzy masks allow taking into account brightness characteristics of water and land in different regions of the Earth at different times of day and year. The results of experimental studies confirm the efficiency of the offered approach to image georeferencing based on the fuzzy masks.

1 Introduction

Inaccurate measurement of satellite angular orientation leads to significant errors in image georeferencing. It is especially typical of geostationary satellites having a great height of an orbit. To reduce these errors, it is necessary to coregister objects in the image and their corresponding referential objects.

In geometric processing of images from Russian geostationary satellite "Elektro-L" coastlines extracted from electronic maps are used as reference objects for precise georeferencing [1].

Referential coastlines are represented as binary masks, where land is coded by unit and water is coded by zero. An example of a binary mask is shown in Fig. 1. Pixels corresponding to water and land are shown in black and white respectively.

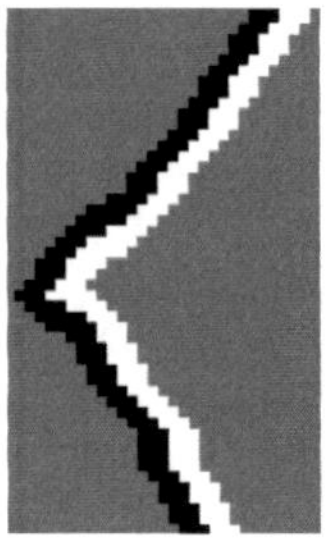

Fig. 1. An example of binary mask.

Coregistration of the image fragment and the binary mask is performed by maximizing the correlation function:

$$\rho = \frac{\overline{Y_1} - \overline{Y_0}}{\sqrt{D}} \cdot \frac{\sqrt{q_1\, q_0}}{q_1 + q_0}, \qquad (1)$$

where $\overline{Y_1}$ and $\overline{Y_0}$ are average values of intensity in pixels under unit and zero parts of the binary mask, D is variance of intensity in pixels under the mask, q_1 and q_0 are numbers of unit and zero values in the mask.

The accuracy and reliability of coregistration are negatively affected by such factors as the partial hiding of coastline in the image by clouds, incomplete conformity of coastlines in the image and electronic map. Therefore, it's necessary to develop alternative approaches to precise georeferencing using reference coastlines.

In the recent years, the theory of fuzzy sets is increasingly used in solving various applied problems, including image processing.

For the problem of precise image georeferencing we propose to replace binary masks of coastlines by fuzzy ones. Our purpose is to research the applicability of fuzzy masks, generated according to the shape of coastlines and brightness characteristics of water and land, in the problem of precise georeferencing.

2 Theoretical part

Fuzzy masks were first proposed by F. Russo and G. Ramponi [2]. They are meant to detect objects in the images.

To generate fuzzy mask, objects located in analyzed image fragment are classified. Each class is associated with a fuzzy set, its membership function is defined in the pixel brightness domain. The value of fuzzy set membership function is considered as the degree of truth of fuzzy logic statement "an object of this class is located in that pixel".

Fuzzy mask defines the relative positions of objects of different classes in the image fragment. Membership

* Corresponding author: vitek-vau@yandex.ru

of the image fragment in the fuzzy mask is defined as conjunction of fuzzy logic statements considered above. It should be noted that generally (depending on the problem) a fuzzy mask can have any form and size.

B. Solaiman, R. Fiset and F. Cavayas suggested using fuzzy masks in the problem of road recognition [3]. Herewith, the ideas of the fuzzy set theory are used in many applied problems [4–9].

An example of the fuzzy mask by size of 3×3, describing a curved fragment of the road, is presented in Fig. 2. Pixels corresponding to the classes "Road" and "Surrounding background" are shown in white and gray respectively. The fuzzy sets associated with these classes have membership functions presented in Fig. 3.

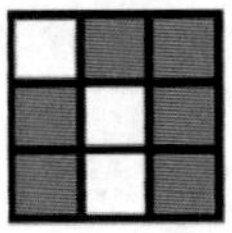

Fig. 2. An example of the fuzzy mask by size 3×3.

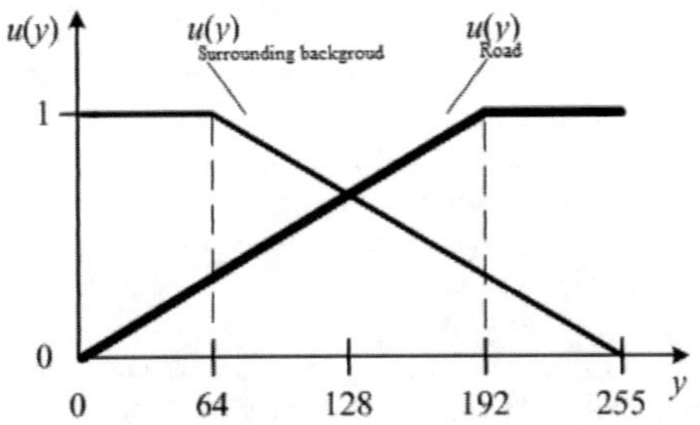

Fig. 3. The membership functions of fuzzy sets "Road" and "Surrounding background" are defined in the pixel brightness domain

To solve the problem of precise image georeferencing using coastlines fuzzy masks it is necessary to introduce the classes "Water" and "Land" into consideration. Thus, the fuzzy mask M defining some coastline consists of n pixels corresponding to one of the two introduced classes.

Membership of the image fragment Y in the fuzzy mask M is defined as

$$u_M(Y) = \bigwedge_{i=1}^{n} u_{c_i}(y_i). \qquad (2)$$

where y_i is the brightness of image pixel under the i th pixel of the fuzzy mask, c_i is the class corresponding to the i th pixel of the fuzzy mask, $u_c(y)$ is the membership function of the fuzzy set associated with the class c, $\wedge$ is the fuzzy conjunction operator.

Fuzzy conjunction may be implemented according to some T-norm.

A T-norm is a function $T:[0,1]\times[0,1]\to[0,1]$, which satisfies the following properties for any $x,y,z,w\in[0,1]$:

- commutativity: $T(x,y)=T(y,x)$;
- monotonicity: $T(x,y)\le T(z,w)$ if $x\le z$ and $y\le w$;
- associativity: $T(x,T(y,z))=T(T(x,y),z)$.

Besides, the boundary conditions are satisfied: $T(0,0)=T(0,1)=T(1,0)=0$ and $T(x,1)=x$ ($\forall x\in[0,1]$).

The following T-norms are most widespread:

$T_1(x,y)=min(x,y)$ and $T_2(x,y)=xy$

($(x,y)\in[0,1]\times[0,1]$).

According to these T-norms, a fuzzy conjunction operator may be defined as

$$u_{A\cap B}(y)=u_A(y)\wedge u_B(y)=\min(u_A(y),u_B(y)) \qquad (3)$$

or

$$u_{A\cap B}(y)=u_A(y)\wedge u_B(y)=u_A(y)u_B(y), \qquad (4)$$

where A and B are the fuzzy sets with the membership functions $u_A(y)$ and $u_B(y)$ respectively.

The expression (3) was used by B. Solaiman, R. Fiset and F. Cavayas to evaluate membership of image fragment in a fuzzy mask under solving the problem of road recognition. However, the use of the expression (4) is more preferable because it's steadier against the noise.

For example, $(n-1)$ pixels of the image fragment have great (close to unit) values of $u_{c_i}(y_i)$ and one pixel contains the noise ($u_{c_i}(y_i)$ is close to zero).

If the expression (3) is used, all $(n-1)$ "good" pixels are ignored and this leads to an erroneous conclusion that the image fragment doesn't correspond to the mask.

At the same time, the use of the expression (4) allows considering all n values of $u_{c_i}(y_i)$. If most of them are close to unit it's concluded that the image fragment corresponds to the mask.

It also requires all values of $u_{c_i}(y_i)$ to be equal to zero. Thus, the membership function of each fuzzy set should be specified so that its minimum value is greater than zero. The minimum value of the membership function $u_{c_i}(y_i)$ is chosen in accordance with the probability of having noise in image pixel.

Taking use of the expression (4) into account, membership of the image fragment in the fuzzy mask can be rewritten as

$$u_M(Y)=\prod_{i=1}^{n} u_{c_i}(y_i). \qquad (5)$$

The flaw in the expression (5) is that value of $u_M(Y)$ is very close to zero in the case of great n and unideal compliance of the image fragment and the fuzzy mask. In such situation the result underflow can arise in machine calculations of the expression (5).

It is expedient to replace multiplication in the expression (5) with the geometric mean:

$$u_M(Y)=\left(\prod_{i=1}^{n} u_{m_i}(y_i)\right)^{\frac{1}{n}}. \qquad (6)$$

The value of $u_M(Y)$ defined as (6) always belongs to the range from $\min_{i=\overline{1,n}}(u_{m_i}(y_i))$ to $\max_{i=\overline{1,n}}(u_{m_i}(y_i))$. Moreover, the result of the comparison of $u_M(Y_1)$ and $u_M(Y_2)$

which are calculated using the expression (6) is the same that in the case of calculation using the expression (5).

Coregistration of the image fragment and reference coastline may be performed by maximizing the function (6).

To manage the coregistration results, it's possible to change membership functions of fuzzy sets "Land" and "Water".

3 The experimental studies

Coastline fragments with various degrees of uniqueness have been used in the experimental studies. The images had various spatial resolution and represent the Earth's surface regions with various areas.

The cases of the partial hiding of coastline in the image by clouds, complete and incomplete conformity of coastline in the image were considered.

Values of intensity in pixels corresponding to water and land depend on geographic location and time of the day and vary from image to image. So, two different pairs of the membership functions of the fuzzy sets "Water" and "Land" shown in Fig. 4 (a, b) have been used in the experimental studies.

The minimum value of all membership functions equal to 0.01 has been chosen, that is the probability of noisy pixels in the images has been estimated to be 0.01.

Coregistration of coastline in the image and reference one is performed by maximizing the correlation function (1) with usage of binary mask, by maximizing the membership function (6) with the usage of a fuzzy mask, and by maximizing the geometric mean of (1) and (6) with the usage of the binary and fuzzy mask combination.

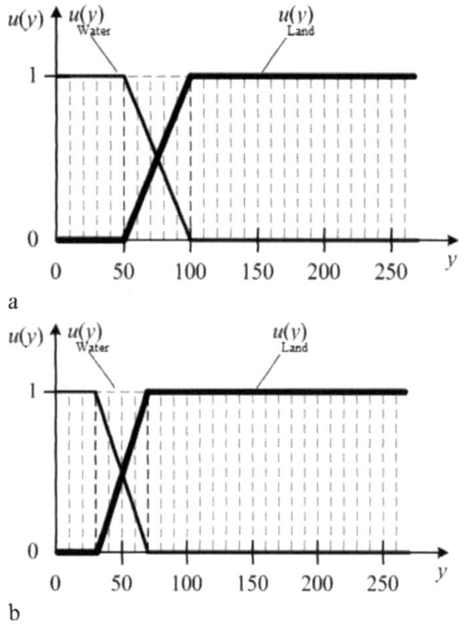

Fig. 4. The examples of the membership functions of the fuzzy sets "Water" and "Land"

Both fuzzy and binary masks provide the exact results at complete conformity of the mask and coastline in the image, good illumination and the lack of clouds.

These types of the masks may provide inaccurate results at low level of the reference coastline uniqueness because the probability of finding other similar borderline between light and dark objects becomes rather high.

Nevertheless, this probability is lower when using the fuzzy masks with thoroughly chosen membership functions of the fuzzy sets "Water" and "Land", than when using the binary masks.

In case of incomplete conformity of the mask and coastline in the image both types of the masks provide rather not exact coregistration.

However, the maximum value of the membership function (6) is rather low in this case, while the maximum value of the correlation function (1) is as high as in case of complete conformity of the mask and coastline in the image. Therefore, it's possible to detect such cases when using fuzzy masks and it's not possible when using binary ones.

Both types of the masks, as a rule, provide incorrect coregistration in case of partial hiding of the coastline by clouds. The maximum values of the functions (1) and (6) are low, that allows detecting such cases and reject incorrect coregistration results. The results of the binary masks are tittle more accurate, however their accuracy is still not high enough.

Thus, the usage of fuzzy masks is preferable with low level of the referential coastline uniqueness and with incomplete conformity of the mask and coastline in the image. In other cases fuzzy and binary masks provide similar results. The combination of fuzzy and binary masks provides, as a rule, more accurate coregistration, than the usage of each mask type separately, but requires more computational resources.

It should be taken into consideration, that some features of georeferencing are based on the fuzzy masks.

The accuracy of georeferencing depends on the choice of the fuzzy sets "Water" and "Land" membership functions.

It's necessary to take into consideration the land brightness characteristics in each region. So, for example, the surface of the Hindustan peninsula is significantly darker in the visible spectral range, than the surface of Africa. Besides, illumination of the land surface depends on time of the day and season. In this regard it is expedient to set individual membership functions to each of these cases, using known aprioristic data.

4 Conclusions

The results of the experiments allowed to make following conclusions. The approach to precise image georeferencing based on coregistration of image fragments and reference coastlines represented as fuzzy masks can be applied in practice.

The use of the formula (6) for membership function's evaluation of the image fragment to the fuzzy mask allows to exclude the possibility of the result overflow. If the restrictions on the performance time of georeferencing are not rigid, it's better to perform the coregistration using the combination of binary and fuzzy masks by maximizing the geometric mean of the functions (1) and (6).

In case of rigid restrictions on the performance time of georeferencing it is possible to use low value of the fuzzy mask membership function as additional rejection criterion of the binary mask and the image fragment coregistration result.

To receive correct results of image georeferencing using fuzzy masks it's necessary first to analyze brightness characteristics of water and land during various times of the day in various regions of the Earth and create database of fuzzy masks membership functions.

It should be noted that fuzzy masks won't have advantage in comparison with binary masks, if the same membership functions are used to all times of the day and all regions of the Earth.

References

1. N.A. Egoshkin, V.V. Eremeev et al., *Current problems in remote sensing of the earth from space*, **6**, 132-138 (2009). [In Rus]
2. F. Russo, G. Ramponi, *Working on Image Data Using Fuzzy Rules, Proc. Sixth European Signal Processing Conf. EUSIPCO-92. Bruxelles*. pp. 1413-1416 (1992)
3. B. Solaiman, R. Fiset, F. Cavayas, *IEEE International Geoscience and Remote Sensing Symposium Proceeding*, pp. 894-896 (1998)
4. L.A. Demidova, N.I. Nesterov, Fuzzy and possibilistic segmentation of earth surface images by means of intelligent information technologies. *Proceedings 2014 International Conference on Computer Technologies in Physical and Engineering Applications, ICCTPEA 2014*, pp. 35–36 (2014)
5. L.A. Demidova, *Automation and Remote Control,* **74,** 313–320 (2013)
6. L.A. Demidova, S. V. Truhanov, *Cloud of Science*, **2,** 40–60 (2015) [In Rus]
7. L. Demidova, Yu. Sokolova, E. Nikulchev, *International Review on Modelling and Simulations*, **8**, 446-457 (2015)
8. L. Demidova, S. Trukhanov, *Contemporary Engineering Sciences*, **8,** 885-900 (2015)
9. D. Van der Weken, M. Nachtegael, E.E. Kerre, *Proceedings of IEEE Int. Conf. Acoustics, Speech and Signal Processing*, pp. 3317-3320 (2002)

The Analytic Solutions to the Task of Maintaining the Specified Characteristics Stability in Aircraft Flight Control System Reconfiguration

Dmitry Efanov[1] and *Vladislav* Kosyanchuk[2,*]

[1]JSC "Spatial information systems", Moscow, Russia
[2]FGUP "GosNIIAS", Moscow, Russia

Abstract. The paper sets out the conceptual issues of methods of analytical solution of the problem of defined aircraft stability characteristics conservation during flight control system reconfiguration under actuating subsystem failure. It is shown that the use of the reconfiguration technique with analytic set of stabilizing control laws synthesized by the decomposition method taking into account the cross-linking between the control channels allows to fully compensate the effects of failures of the actuating subsystem in each of the control channels under the conditions of complete controllability of aircraft as a dynamic system. It thus makes it possible to reconfigure the flight control system that always ensures the stability of the motion of the aircraft with a precisely prescribed poles. During study of failures in the form of a gradual reduction of the effectiveness of control surfaces it found that the synthesized analytical control law is a corresponding automatic increase feedback gain serviceable channel actuating subsystem, as well as the reallocation of control signals between the drives within the existing cross-connections between the control channels. Some of the solutions obtained by reconfiguring the control laws are not satisfactory, because of significantly limited possibility of maneuvering.

1 Introduction

Improving aircraft safety determines the relevance of the development of flight control systems capable of detecting failures and compensating for their effects. Modern aircraft are characterized by a high degree of analytical (functional) redundancy of actuating flight control system subsystems, so in the event of failure it is possible to redistribute functions.

At present, the functional (analytical) reconfiguration of the actuating subsystem has not been fully utilized.

There are passive and active approaches to reconfiguring the flight control system [1–2]. In the passive approach, the aim is to obtain a control law, which meets all the criteria set both in normal and emergency situations. An active approach is based on the failure identification scheme.

Development of analytical methods for reconfiguring the flight control system is a relatively new filed [3–4]. However, these studies do not cover sufficiently the task of ensuring the specified stability characteristics when reconfiguring the flight control system in case of failure and damage of the actuating subsystem elements. Therefore, the results obtained in these studies cannot be fully utilized to solve the reconfiguration problem.

The aim of this work is to find an analytical solution to maintain specified characteristics of the aircraft stability during flight control system reconfiguration.

2 Results and discussions

Consider a linear MIMO (Multi Input Multiple Output) system, the equations of which are defined as follows:

$$\sigma \mathbf{x}(t) = \boldsymbol{A}\mathbf{x}(t) + \boldsymbol{B}\mathbf{u}(t), \ \mathbf{y}(t) = \boldsymbol{C}\mathbf{x}(t). \qquad (1)$$

Where t is time that can be either continuous $t \in \mathbb{R}_+ = [0, +\infty)$, or discrete $t \in \mathbb{N}_0 = \{0,1,2,\ldots\}$. The vector $\mathbf{x} \in \mathbb{R}^n$ corresponds to the dimension state vector n; $\mathbf{u} \in \mathbb{R}^r$ to the dimension input (control) vector r; $\mathbf{y} \in \mathbb{R}^m$ to the output (measurement) vector m. Operator σ means either time differentiation for the continuous system, i.e. $\sigma\mathbf{x}(t) = \dot{\mathbf{x}}(t)$, or time offset operation for the discrete system, i.e. $\sigma\mathbf{x}(t) = \mathbf{x}(t+1)$. The set $\mathbb{R}$ means real numbers (real axis), and $\mathbb{R}^k$ means the dimension real space k accordingly.

It is believed that the set of eigenvalues of matrix $\boldsymbol{A}$

$$\operatorname{eig}(\boldsymbol{A}) = \{\lambda_i \in \mathbb{C} : \det(\lambda \boldsymbol{I}_n - \boldsymbol{A}) = 0\}$$

inevitably includes such complex numbers $\lambda_i \in \mathbb{C}$ that their real parts $\operatorname{Re}(\lambda_i) > 0$ for the case $\sigma\mathbf{x}(t) = \dot{\mathbf{x}}(t)$, or

* Corresponding author: kos.vl.v@gmail.com

their modulus $|\lambda_i| > 1$ for the case $\sigma\mathbf{x}(t) = \mathbf{x}(t+1)$. Here $\boldsymbol{I}_n$ is an identity matrix of order n; $|\lambda_i|$ is an eigenvalue modulus λ_i; $\mathbb{C}$ is a set of complex numbers (complex plane).

We introduce the concept of set $\mathbb{C}^{\text{stab}}$, which represents, depending on the type of the dynamic system (continuous or discrete one) respectively left half-plane $\mathbb{C}^-$, i.e. $\mathbb{C}^{\text{stab}} \doteq \mathbb{C}^-$, or the area within a circle of unit radius centered at the beginning of $\mathbb{C}$, i.e. $\mathbb{C}^{\text{stab}} \doteq \mathbb{C}_{|\lambda|<1}$. Here $\text{eig}(\boldsymbol{A}) \subset \mathbb{C}^{\text{stab}}$ is a condition of asymptotic stability expressed in terms of eigenvalues (poles) of the matrix.

It is assumed that for the MIMO-system (1) some control exists with a negative feedback of type:

$$\mathbf{u}(t) = -\boldsymbol{K}\mathbf{x}(t), \tag{2}$$

where $\boldsymbol{K} \in \mathbb{R}^{r\times n}$ is a matrix of coefficients (regulator matrix).

Assume that $\boldsymbol{B}^{\perp}$ is a zero divisor of the matrix meeting the following conditions: $\boldsymbol{B}^{\perp}\boldsymbol{B} = \boldsymbol{0}_{(n-r)\times r}$, $\boldsymbol{B}^{\perp}\boldsymbol{B}^{\perp\mathrm{T}} = \boldsymbol{I}_{n-r}$, where $\boldsymbol{B}^{+}$ is the Moore-Penrose pseudoinverse; $\boldsymbol{0}_{r\times r}$ is a zero matrix of size $r \times r$.

Consider the following multi-level decomposition of the dynamical system (1) with matrices $\boldsymbol{A} \in \mathbb{R}^{n\times n}$, $\boldsymbol{B} \in \mathbb{R}^{n\times m}$ and $\boldsymbol{C} = \boldsymbol{I}_n$ [5].

The zero (original) level is

$$\boldsymbol{A}_0 = \boldsymbol{A},\ \boldsymbol{B}_0 = \boldsymbol{B}, \tag{3}$$

the first level is

$$\boldsymbol{A}_1 = \boldsymbol{B}^{\perp}\boldsymbol{A}\boldsymbol{B}^{\perp\mathrm{T}},\ \boldsymbol{B}_1 = \boldsymbol{B}^{\perp}\boldsymbol{A}\boldsymbol{B}, \tag{4}$$

kth (intermediate) level is

$$\boldsymbol{A}_k = \boldsymbol{B}_{k-1}^{\perp}\boldsymbol{A}_{k-1}\boldsymbol{B}_{k-1}^{\perp\mathrm{T}},\ \boldsymbol{B}_k = \boldsymbol{B}_{k-1}^{\perp}\boldsymbol{A}_{k-1}\boldsymbol{B}_{k-1}, \tag{5}$$

Lth (final) level is $L = \text{ceil}(n/r) - 1$,

$$\boldsymbol{A}_L = \boldsymbol{B}_{L-1}^{\perp}\boldsymbol{A}_{L-1}\boldsymbol{B}_{L-1}^{\perp\mathrm{T}},\ \boldsymbol{B}_L = \boldsymbol{B}_{L-1}^{\perp}\boldsymbol{A}_{L-1}\boldsymbol{B}_{L-1}. \tag{6}$$

Here $\text{ceil}(*)$ is rounding operation for numbers * to the greater value, e.g., $\text{ceil}(0.3) = 1$, $\text{ceil}(1.1) = 2$, $\text{ceil}(2.01) = 3$, etc.

If the MIMO system (1) with two matrices $(\boldsymbol{A}, \boldsymbol{B})$ is fully controllable, then all systems with matrices $(\boldsymbol{A}_i, \boldsymbol{B}_i)$ (4) – (5) are controllable.

Assume that the MIMO system (1) is fully controllable and all matrices $\boldsymbol{B}_i$ have full ranks in rows, and matrix $\boldsymbol{K} \in \mathbb{R}^{r\times n}$ meets "assembly" formulas:

$$\boldsymbol{K} = \boldsymbol{K}_0 = \boldsymbol{B}_0^{-}\boldsymbol{A} - \Phi_0\boldsymbol{B}_0^{-},\quad \boldsymbol{B}_0^{-} = \boldsymbol{B}_0^{+} + \boldsymbol{K}_1\boldsymbol{B}_0^{\perp}, \tag{7}$$

$$\boldsymbol{K}_1 = \boldsymbol{B}_1^{-}\boldsymbol{A}_1 - \Phi_1\boldsymbol{B}_1^{-},\quad \boldsymbol{B}_1^{-} = \boldsymbol{B}_1^{+} + \boldsymbol{K}_2\boldsymbol{B}_1^{\perp}, \ldots \tag{8}$$

$$\boldsymbol{K}_k = \boldsymbol{B}_k^{-}\boldsymbol{A}_k - \Phi_k\boldsymbol{B}_k^{-},\quad \boldsymbol{B}_k^{-} = \boldsymbol{B}_k^{+} + \boldsymbol{K}_{k+1}\boldsymbol{B}_k^{\perp}, \ldots \tag{9}$$

$$\boldsymbol{K}_L = \boldsymbol{B}_L^{-}\boldsymbol{A}_L - \Phi_L\boldsymbol{B}_L^{-}, \tag{10}$$

Then the following equation is correct:

$$\text{eig}(\boldsymbol{A} - \boldsymbol{B}\boldsymbol{K}) = \bigcup_{i=1}^{L+1}\text{eig}(\Phi_{i-1}). \tag{11}$$

The limitation of rank completeness for matrices $\boldsymbol{B}_i$ is in contrast with the presence of functional redundancy of the flight control system; therefore an algorithm has been designed to eliminate this limitation. It comprises the following steps:

1) check the rank of matrix $\boldsymbol{B}$;
2) if $\text{rank}\,\boldsymbol{B} < r$, then perform a skeleton decomposition $\boldsymbol{B} = \widehat{\boldsymbol{B}}\boldsymbol{T}$;
3) define matrices $\boldsymbol{A}_0 = \boldsymbol{A}$, $\boldsymbol{B}_0 = \widehat{\boldsymbol{B}}$;
4) calculate $L = \text{ceil}(n/r) - 1$;
5) set matrices $\Phi = \Phi_0$, Φ_1, …, Φ_L, such that $\bigcup_{i=1}^{L+1}\text{eig}(\Phi_{i-1})$ a desired set of eigenvalues of the closed system;
6) calculate the zero divisor $\boldsymbol{B}_0^{\perp}$, and then matrices $\boldsymbol{A}_1 = \boldsymbol{B}_0^{\perp}\boldsymbol{A}_0\boldsymbol{B}_0^{\perp+}$, $\boldsymbol{B}_1 = \boldsymbol{B}_0^{\perp}\boldsymbol{A}_0\boldsymbol{B}_0, \ldots$, $\boldsymbol{A}_k = \boldsymbol{B}_{k-1}^{\perp}\boldsymbol{A}_{k-1}\boldsymbol{B}_{k-1}^{\perp+}$, $\boldsymbol{B}_k = \boldsymbol{B}_{k-1}^{\perp}\boldsymbol{A}_{k-1}\boldsymbol{B}_{k-1}$;
7) calculate the zero divisor $\boldsymbol{B}_k^{\perp}$, and then matrices $\boldsymbol{A}_{k+1} = \boldsymbol{B}_k^{\perp}\boldsymbol{A}_k\boldsymbol{B}_k^{\perp+}$, $\boldsymbol{B}_{k+1} = \boldsymbol{B}_k^{\perp}\boldsymbol{A}_k\boldsymbol{B}_k, \ldots$;
8) determine the zero divisor $\boldsymbol{B}_{L-2}^{\perp}$, and then matrices $\boldsymbol{A}_{L-1} = \boldsymbol{B}_{L-2}^{\perp}\boldsymbol{A}_{L-2}\boldsymbol{B}_{L-2}^{\perp+}$, $\boldsymbol{B}_{L-1} = \boldsymbol{B}_{L-2}^{\perp}\boldsymbol{A}_{L-2}\boldsymbol{B}_{L-2}$;
9) calculate the zero divisor $\boldsymbol{B}_{L-1}^{\perp}$, and then matrices $\boldsymbol{A}_L = \boldsymbol{B}_{L-1}^{\perp}\boldsymbol{A}_{L-1}\boldsymbol{B}_{L-1}^{\perp+}$, $\boldsymbol{B}_L = \boldsymbol{B}_{L-1}^{\perp}\boldsymbol{A}_{L-1}\boldsymbol{B}_{L-1}$;
10) successively calculate matrices

$$\boldsymbol{K}_L = \boldsymbol{B}_L^{-}\boldsymbol{A}_L - \Phi_L\boldsymbol{B}_L^{-},$$

$$\boldsymbol{B}_{L-1}^{-} = \boldsymbol{K}_L\boldsymbol{B}_{L-1}^{\perp} + \boldsymbol{B}_{L-1}^{+},\ \boldsymbol{K}_{L-1} = \boldsymbol{B}_{L-1}^{-}\boldsymbol{A}_{L-1} - \Phi_{L-1}\boldsymbol{B}_{L-1}^{-}, \ldots,$$

$$\boldsymbol{B}_k^{-} = \boldsymbol{K}_{k+1}\boldsymbol{B}_k^{\perp} + \boldsymbol{B}_k^{+},\ \boldsymbol{K}_k = \boldsymbol{B}_k^{-}\boldsymbol{A}_k - \Phi_k\boldsymbol{B}_k^{-}, \ldots,$$

$$\boldsymbol{B}_1^{-} = \boldsymbol{K}_2\boldsymbol{B}_1^{\perp} + \boldsymbol{B}_1^{+},\ \boldsymbol{K}_1 = \boldsymbol{B}_1^{-}\boldsymbol{A}_1 - \Phi_1\boldsymbol{B}_1^{-},$$

$$\boldsymbol{B}_0^{-} = \boldsymbol{K}_1\boldsymbol{B}_0^{\perp} + \boldsymbol{B}_0^{+},\ \boldsymbol{K} = \boldsymbol{T}^{+}\boldsymbol{K}_0 = \boldsymbol{T}^{+}\left(\boldsymbol{B}_0^{-}\boldsymbol{A}_0 - \Phi_0\boldsymbol{B}_0^{-}\right).$$

The obtained regulator

$$\boldsymbol{K} = \boldsymbol{T}^{+}\boldsymbol{K}_0 = \boldsymbol{T}^{+}\left(\boldsymbol{B}_0^{-}\boldsymbol{A}_0 - \Phi_0\boldsymbol{B}_0^{-}\right),\ \boldsymbol{B}_0^{-} = \boldsymbol{K}_1\boldsymbol{B}_0^{\perp} + \boldsymbol{B}_0^{+} \tag{12}$$

ensures fulfilment of condition (11), that is the condition of specified pole placement.

Based on the analysis of the relationships in the discussed decomposition method of pole placement, sources of equivalent control laws are identified. There are three sources of equivalent solutions in the form of control laws, i.e. a set of $\boldsymbol{K}_{\{\}}$ regulator matrices, for which the following equation is satisfied

$$\text{eig}\left(\boldsymbol{A} - \boldsymbol{B}\boldsymbol{K}_{\{\}}\right) = \bigcup_{i=1}^{L+1}\text{eig}(\Phi_{i-1}). \tag{13}$$

1. From the guidelines of the above method and the pole placement algorithm constructed on its basis, it follows that, in general, no restriction is imposed on matrices Φ_{i-1} except the conformity of their sets of values with the predetermined values.

2. Another source for a set of equivalent laws $\boldsymbol{K}_{\{\}}$ is the loss of the full rank in the columns of matrices $\boldsymbol{B}_i$ in the decomposition process (3) – (6).

Consider the "skeleton" decomposition $\boldsymbol{B}_i = \widehat{\boldsymbol{B}}_i\boldsymbol{T}_i$ in case of loss of the full rank in the columns of matrix $\boldsymbol{B}_i$. In this case, the equation of sets of eigenvalues is satisfied

$$\mathrm{eig}\left(A_i - \hat{B}_i\hat{K}_i\right) = \mathrm{eig}\left(A_i - B_i T_i^+ \hat{K}_i\right),$$

where $T_i T_i^+ = I$. If, according to formula (12) matrix $T_i^+ \hat{K}_i$ is completed with a summand of the zero divisor, $B_{iR}^{\perp}\varpi_i$, where $B_{iR}^{\perp}$ is the right zero divisor of matrix B_i, i.e. $B_i B_{iR}^{\perp} = 0$, ϖ_i is a matrix of appropriate size with arbitrary elements, then the following equation will be satisfied

$$\mathrm{eig}\left(A_i - \hat{B}_i\hat{K}_i\right) = \mathrm{eig}\left(A_i - B_i\left(T_i^+\hat{K}_i + B_{iR}^{\perp}\varpi\right)\right). \quad (14)$$

Thus, the addition to regulators $\hat{K}_i$ obtained at each level of decomposition of the mathematical model of the MIMO system (1) generates a proper subset of the set of equivalent control laws.

3. Finally, the third source generating the equivalent laws is an incomplete rank in the columns of matrix $B = B_0$ of the model (1). This source from a formal point of view falls under the above discussions about the "skeleton" decomposition of $B_i = \hat{B}_i T_i$. However, it is placed as a separate category because only addition $B_{0R}^{\perp}\varpi_i = B_R^{\perp}\varpi_i$ (at the zero level of decomposition) is additive and all the others can be regarded as multiplicative due to the "assembly" procedures of the regulator (7) – (10).

The mid-range passenger plane considered in the paper is made by the normal scheme. The arrow wing is low and high-lift (slats and single-slotted flaps). The vertical tail is single-fin; the horizontal stabilizer is located on the rear fuselage. Aerodynamic pitch is controlled by a stabilizer (balancing) and elevators, roll is controlled by ailerons, and yaw is controlled by a rudder.

To be definite in solving problems of reconfiguring the flight control system, we will use a continuous linear model of aircraft motion, i.e. in this case $\sigma\mathbf{x}(t) = \dot{\mathbf{x}}(t)$. The state vector of the aircraft model (1) is as follows

$$\mathbf{x} = \begin{pmatrix} V & \alpha & \Theta & \omega_z & \beta & \omega_x & \omega_y & \gamma \end{pmatrix}^{\mathrm{T}} \in \mathbb{R}^8,$$

where $\omega_x, \omega_y, \omega_z$ are angular velocities of roll, yaw and pitch, respectively; $\alpha, \Theta, \gamma, \beta$ are angles of attack, flight path, roll and glide, respectively; V is flight velocity. The aircraft model control vector (1) is

$$\mathbf{u} = \begin{pmatrix} \delta_{\text{elevator}}^{\text{left}} & \delta_{\text{elevator}}^{\text{right}} & \delta_{\text{rudder}} & \delta_{\text{aileron}}^{\text{left}} & \delta_{\text{aileron}}^{\text{right}} \end{pmatrix}^{\mathrm{T}} \in \mathbb{R}^5,$$

where $\delta_{\text{elevator}}^{\text{left}}$, $\delta_{\text{elevator}}^{\text{right}}$ are deflection angles of the left and right elevators, respectively, with deflection ranges of (–27°…22°); δ_{rudder} is rudder deflection angle (–27°…27°); $\delta_{\text{aileron}}^{\text{left}}, \delta_{\text{aileron}}^{\text{right}}$ angles of the left and right ailerons, respectively, with deflection ranges of (–27°…15°).

Using the methodology of dividing the motion into disturbance and non-disturbance motion, the matrix representation of longitudinal and lateral motion of the aircraft has the following form

$$A = \begin{pmatrix} a_{11} & a_{12} & a_{13} & 0 & 0 & 0 & 0 & 0 \\ a_{21} & a_{22} & a_{23} & 1 & 0 & 0 & 0 & 0 \\ -a_{21} & -a_{22} & -a_{23} & 0 & 0 & 0 & 0 & 0 \\ a_{41} & a_{42} & a_{43} & a_{44} & 0 & 0 & 0 & 0 \\ 0 & 0 & 0 & 0 & a_{55} & a_{56} & a_{57} & a_{58} \\ 0 & 0 & 0 & 0 & a_{65} & a_{66} & a_{67} & 0 \\ 0 & 0 & 0 & 0 & a_{75} & a_{76} & a_{77} & 0 \\ 0 & 0 & 0 & 0 & 0 & 1 & a_{87} & 0 \end{pmatrix},$$

$$B = \begin{pmatrix} 0 & 0 & 0 & 0 & 0 \\ b_{21} & b_{22} & 0 & b_{25} & b_{26} \\ -b_{21} & -b_{22} & 0 & -b_{25} & -b_{26} \\ b_{41} & b_{42} & 0 & b_{45} & b_{46} \\ 0 & 0 & b_{54} & 0 & 0 \\ b_{61} & -b_{62} & b_{64} & b_{65} & -b_{66} \\ b_{71} & -b_{72} & b_{74} & b_{75} & -b_{76} \\ 0 & 0 & 0 & 0 & 0 \end{pmatrix}. \quad (15)$$

As can be seen from the structure of the control efficiency matrix B, this aircraft model allows for cross couplings between control channels (controls) of the aircraft for longitudinal and lateral motion.

Since the change of the aircraft flight mode effects aerodynamic coefficients a_{ij} and b_{ij}, it is assumed that failures and damage of the actuating subsystem of the aircraft flight control system also lead to changes in coefficients b_{ij}, and in case of full loss of control they go down to zero. The worst cases of failures in the channel of the elevator and ailerons of the actuating subsystem are associated with a complete violation of the mutual "symmetry" of coefficients in the matrix B, with which inequalities $b_{i1} \neq b_{i2}$, $b_{i5} \neq b_{i6}$, $i \in \{2,4,6,7\}$ are satisfied.

In this paper, the problem of reconfiguring the flight control system was solved analytically, i.e., without numerical errors. To this end, two approaches to solving this problem were considered.

1. Sources are analyzed that generate a set of equivalent control laws $K_{\{\}}$, namely, 1) freedom to choose (set) matrices Φ_{i-1} in (7) – (10); 2) incomplete rank in columns of matrix B_i in the decomposition procedure (3) – (6); 3) incomplete rank in columns of matrix $B = B_0$ in the initial model of the MIMO system (1). In the latter case, the control law includes explicitly an additive component of $B_R^{\perp}\varpi$.

To obtain calculated ratios of analytical reconfiguration of the flight control system, it is believed that an operating object and an object with a failure of the actuating subsystem are presented by corresponding linear MIMO-systems of type (1):

$$\sigma\mathbf{x}(t) = A\mathbf{x}(t) + B\mathbf{u}(t),\ \mathbf{y}(t) = \mathbf{x}(t),\ C = I_n,$$

$$\sigma\mathbf{x}_{\text{failure}}(t) = A\mathbf{x}_{\text{failure}}(t) + B_{\text{failure}}\mathbf{u}_{\text{failure}}(t),$$

$$\mathbf{y}_{\text{failure}}(t) = \mathbf{x}_{\text{failure}}(t),\ C_{\text{failure}} = I_n.$$

DOI: 10.1051/matecconf/20179903004

Thus, the failure of the actuating subsystem of the flight control system is modeled by matrix $\boldsymbol{B}_{\text{failure}}$. Here all variables of the operating and failed objects are observed. It is also believed that for matrix $\boldsymbol{B}$ the rank of the columns can be equal to or less than the number of columns r, which in turn is no less than the dimension of the state space, i.e. $\text{rank}\,\boldsymbol{B} \le r < n$.

Control laws for the properly operating aircraft and the flight control system of the reconfigured object correspond to the following equations, respectively, $\mathbf{u}(t) = -\boldsymbol{K}\mathbf{x}(t)$, $\mathbf{u}_{\text{failure}}(t) = -\boldsymbol{K}_{\text{rec}}\mathbf{x}_{\text{failure}}(t)$, where $\boldsymbol{K}_{\text{rec}}$ is a matrix of the reconfigured control law.

It is believed that the regulator matrix in the control law for the properly operating flight control system is given by (7)

$$\boldsymbol{K} = \boldsymbol{B}_0^- \boldsymbol{A} - \Phi \boldsymbol{B}_0^- + \boldsymbol{B}_R^\perp \varpi, \quad \boldsymbol{B}_0^- = \boldsymbol{B}_0^+ + \boldsymbol{K}_1 \boldsymbol{B}_0^\perp,$$

and the matrix of reconfigured regulator is $\boldsymbol{K}_{\text{rec}} \in \boldsymbol{K}_{\{\}}$. Then the problem of reconfiguration will be to ensure equality of eigenvalues sets:

$$\text{eig}\left(\boldsymbol{A} - \boldsymbol{B}_{\text{failure}}\boldsymbol{K}_{\{\bullet\}}\right) = \text{eig}\left(\boldsymbol{A} - \boldsymbol{B}\boldsymbol{K}\right),$$

where $\boldsymbol{K}_{\text{rec}} = \boldsymbol{K}_{\{\bullet\}}$ is an appropriate element of the analytic set of control laws $\boldsymbol{K}_{\{\}}$ providing the desired placement of the poles. Here the class condition (belonging to the considered regulator set of the properly operating flight control system) is also met, i.e.

$$\left.\begin{array}{r} \boldsymbol{K} = \boldsymbol{B}_0^- \boldsymbol{A} - \Phi \boldsymbol{B}_0^- + \boldsymbol{B}_R^\perp \varpi \\ \boldsymbol{B}_0^- = \boldsymbol{B}_0^+ + \boldsymbol{K}_1 \boldsymbol{B}_0^\perp \end{array}\right\} \in \boldsymbol{K}_{\{\}} .$$

2. Another approach to reconfiguring the flight control system is an analytical synthesis of a set of control laws $\boldsymbol{K}_{\{\}}$ for the “worst” (in some sense) case of failure of the actuating subsystem. The control law under normal functioning of the flight control system, as well as all others failures less dangerous for flight safety will be elements of this set of control laws.

3 Conclusions

Thus, the main provisions of the analytical solution are formulated for the problem of maintaining the specified aircraft stability characteristics in case of the flight control system reconfiguration due to failures and malfunctions of the actuating subsystem. The singularity of these results is that implementation and adaptation of the developed algorithms in on-board control complexes of modern and advanced aircraft will ensure a high level of flying qualities for states of failure to provide safe continuation and completion of the flight.

It is shown that the decomposition method of pole placement used in the synthesis of stabilizing control laws can be applied to reconfigure the flight control system. The analytical synthesis of control laws was demonstrated, as well as the procedure for constructing a set of equivalent solutions on the basis of identified sources of ambiguity (freedom of choice).

References

1. T. A. Johansen, T. I. Fossen, *Automatica*, **49**(5), 1087–1103 (2013)
2. X. Yu, J. Jiang, *Annual Reviews in Control*, **39**, 46–57 (2015)
3. E. Yu. Zybin, V. V. Kosyanchuk, *Journal of Computer and Systems Sciences International*, **49**(1), 105–114 (2010)
4. E. Yu. Zybin, V. V. Kosyanchuk, A. M. Kulchak, *Mekhatronika, Avtomatizatsiya, Upravlenie*, **7**, 59–66 (2014)
5. E.A. Mikrin, V.N. Ryabchenko, N.E. Zubov, *Matrix methods in the theory and practice of automatic control systems of aircraft* (Moscow, Publishing House of N.E. Bauman, 2016)

DOI: 10.1051/matecconf/20179903005

Integration of Predictive Display and Aircraft Flight Control System

A.V. Efremov[1,]*, *M.S.* Tjaglik[1], *I.H.* Irgaleev[1] and *V.G.* Tsipenko[2]

[1]Moscow Aviation Institute, Aeronautical school, Moscow, Russia
[2]Moscow State Technical University of Civil Aviation, Moscow, Russia

Abstract. The synthesis of predictive display information and direct lift control system are considered for the path control tracking tasks (in particular landing task). The both solutions are based on pilot-vehicle system analysis and requirements to provide the highest accuracy and lowest pilot workload. The investigation was carried out for cases with and without time delay in aircraft dynamics. The efficiency of the both ways for the flying qualities improvement and their integration is tested by ground based simulation.

1 Introduction

The improvement of flying qualities might be realized by the different ways. One of them is the usage of direct lift control (DLC) system providing the decoupling of path and angular motions. It allows to decrease the pole order in the origin in aircraft path motion dynamics [1, 2] and to get the slope of its amplitude frequency response characteristics close to –20 dB/dec in crossover frequency range. The other way is the usage of predictive display [3]. It influences on zeros of controlled element dynamics transfer function and allows to get the same slope of its amplitude frequency characteristic (-20 dB/dec) in the same frequency range. The high potentiality of predictive display was investigated in flight tests [4] and in ground-based simulation [5]. In current paper it is considered two approaches to the design of predictive display differed by the way of determining the path angle. The effectiveness of the predictive display compares with effectiveness of DLC system. The last one was realized by two ways: by use command filters and by use the feedback and feed forward controllers. The additional requirement to the dynamics of DLC aircraft based on requirement of closeness of its frequency response characteristic to the optimal control dynamics [6] is considered. The efficiency of such additional requirement was checked in ground-based simulation. The potentialities of integration of DLC system and predictive display was investigated in ground based simulation.

2 The usage of direct lift control (DLC) system

A number of studies were dedicated to the usage of DLC system for flying qualities improvement [1, 2, 7, 8] where the requirement to execution of a specific mode of motion was formulated.

In the current investigation two versions of DLC system providing the variable angle of attack and constant pitch angle are considered. These versions are shown on fig 1-2. The version given on fig 1 provides the required mode of motion with help of command filters W_1 and W_2. The second version is based on the usage of the command filter in the angle of attack loop and feed forward controller in the other loop filter fig. 2. Two actuator dynamics with and without rate limit were investigated.

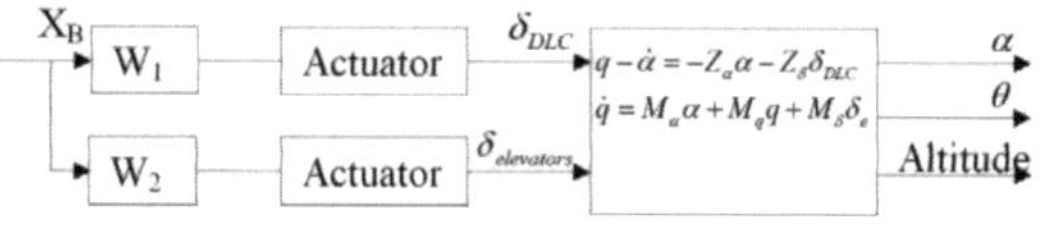

Fig 1. DLC system with command filters

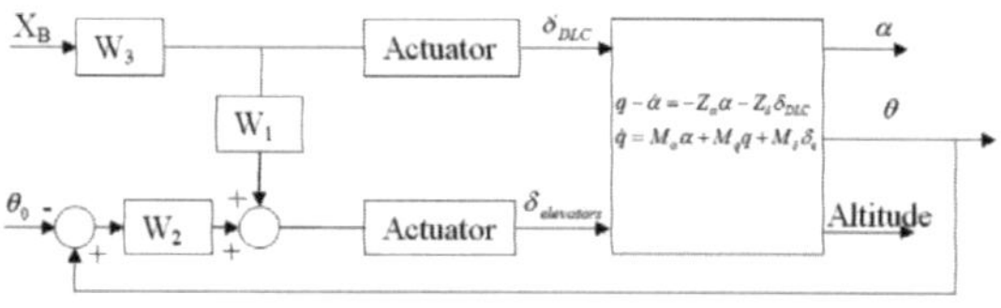

Fig 2. DLC system with command filters and feedback and feed forward controllers

The equations for the filters and controllers were defined by using the equations of short period motion. For any versions of the considered DLC systems the use of DLC leads to the decrease of pole order in the origin of the controlled element dynamics (table 1). It provides the simplification of pilot behavior (the decrease of his

* Corresponding author: pvl@mai.ru

lead compensation). As a consequence, the accuracy in tracking task has to be improved too.

Preliminary experiments were executed on the MAI workstation where the operator carried out the single loop tracking task with the input signal $i(t)$ characterized by the low bandwidth spectrum $S_{ii} = K^2/(\omega^2+\omega_i^2)^2$, where $\omega_i = 0.2$ rad/sec.

Table 1. W_C for different versions of DLC system

	version 1	version 2
W1	K_C	$\frac{W_3\cdot(-Z_\delta)\cdot M_\alpha}{M_\delta\cdot(p-Z_\alpha)}$
W2	$\frac{W_1\cdot(-Z_\delta)\cdot M_\alpha}{M_\delta\cdot(p-Z_\alpha)}$	K_C
W3	---	K_C
Wc= $\frac{H}{X_e}$	$\frac{V\cdot(-Z_\delta)}{p\cdot(p-Z_\alpha)}$	$\frac{V\cdot(-Z_\delta)}{p\cdot(p-Z_\alpha)}$

The measurement of the accuracy demonstrated that the close results were achieved for two versions of DLC system (fig3). In case of «ideal» actuator (when the rate limit was not tacken into account) the variance of error was three times higher in comparison with experiments carried out for aircraft without DLC system.

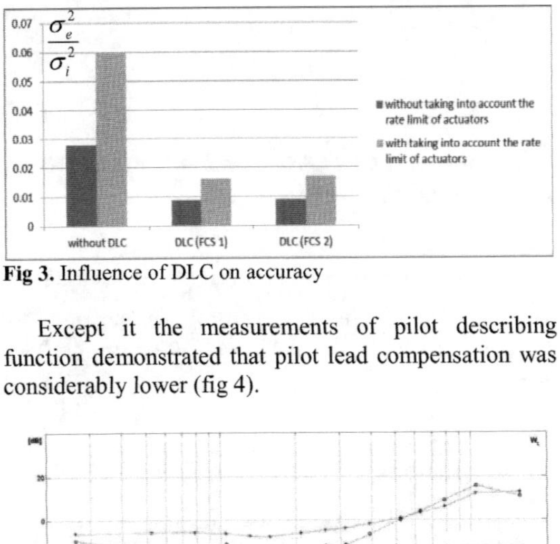

Fig 3. Influence of DLC on accuracy

Except it the measurements of pilot describing function demonstrated that pilot lead compensation was considerably lower (fig 4).

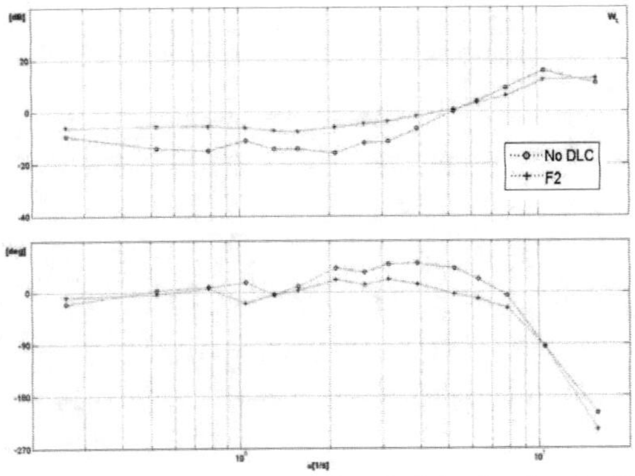

Fig 4. Pilot frequency response characteristics for aircraft with and without DLC

In case when the rate limit was taken into account in the model of actuator dynamics, the variance of error was higher although the effect of DLC increased up to 4-4.5 times. Because the parameters of filters are defined by the aerodynamic coefficients (see table 1), their inaccurate knowledge does not allow to keep the required mode of motion. The evaluation of influence of such inaccurcy on the required mode of motion was examined on simulator. The experiments demonstrated that inaccuracy of the knowledge of coefficients $(M_\alpha, Z_\alpha, Z_\delta)$ up to 30% didn't have any practical influence on variance of error.

Except the requirement to the mode of motion, an additional one – the closeness of DLC aircraft dynamic to the optimal controlled element dynamics (W_C^{opt}), was added.

The optimal control dynamics was defined in [6,9] as the transfer function provided the highest accuracy $\sigma_e^2 = \sigma_{emin}^2$ with the simplest type of pilot behavior when his describing function corresponds to $W_P(jw) = K_P e^{-jw\tau}$, and spectral density of pilot noise $n_e(t)$ ("remnant") $S_{n_e n_e} = K_{n_e}\sigma_{emin}^2$, K_{n_e}=0,01. The Wiener approach was used for the evaluation with taking into account the fixed part - non minimum phase element $e^{-s\tau}$. According to [6,9], the general equation for the W_C^{opt} obtained by use of Wiener approach is the following:

$$W_C^{opt}(s) = \frac{M_0(s)\hat{P}^-(s)}{A^+(s)\hat{P}^-(s) - M_0(s)P^-(s)},$$

where $P^-(s)$ and $\hat{P}^-(s)$ are the nominator and denominator of the time delay Pade approximation.

$$W_O(s) = e^{-s\tau} \cong \frac{2-s\tau}{2+s\tau} = \frac{P^-(s)}{\hat{P}^-(s)}.$$

The polynomials $A^*(s)$ and $C(s)$ can be defined by factorization of equation for the input signal spectral density:

$$S_{ii}(s) = \frac{C(s)}{D(s)} = \frac{C^+(s)C^-(s)}{D^+(s)D^-(s)}$$

and spectral density:

$$S(s) = S_{ii}(s) + S_{n_e n_e}(s) = \frac{A(s)}{B(s)} = \frac{A^+(s)A^-(s)}{B^+(s)B^-(s)}.$$

The polynomial $M_0(s)$ is defined by equation

$$P^-(s)M_0(s)A^-(s) + B^+(s)L_0(s) = C(s)\hat{P}^-(s). \quad (1)$$

With taking into account the remnant spectral density model given above it is possible to get that $B(s) = D(s)$ and equation for $W_C^{opt}(s)$:

$$W_C^{opt}(s) = \frac{A^-(s)M_0(s)\hat{P}^-(s)}{D^+(s)[N^2D^-(s)\hat{P}^-(s) + L_0(s)]},$$

where $N^2 = S_{n_e n_e}$.

It is shown in [8] that

$$N^2 = S_{n_e n_e} = \frac{\sigma^2_{e\,min}}{\frac{1}{K_{n_e}} - \int_0^\infty |W_{CL}|^2 d\omega}.$$

For the case when $S_{ii} = \frac{K^2}{(\omega^2 + \omega_i^2)^2}$:

$$A^+(s) = \sqrt{K_{n_e}\sigma^2_{e\,min}(s^2 + bs + c)};$$

$$A^-(s) = \sqrt{K_{n_e}\sigma^2_{e\,min}(s^2 - bs + c)};$$

where $b = \sqrt{2(\omega_i^2 + c)}$, $c = \sqrt{K_i^2 / (K_{n_e}\sigma^2_{e\,min}) + \omega_i^4}$, $B^+(s) = (\omega_i + s)^2$, $C_p = K_i^2$.

These equations allows transforming (1) to the following:

$$\begin{aligned}(2 - s\tau)M_0(s)\sqrt{K_{n_e}\sigma^2_{e\,min}(s^2 - bs + c)} + \\ + (\omega_i + s)^2 L_0(s) = K_i^2(2 + s\tau)\end{aligned} \quad (2)$$

The solution of this equation is the following:

$$M_0(s) = a_0 + b_0 s,\ L_0(s) = c_0 + d_0 s + f_0 s^2$$

The coefficients a_0, b_0, c_0, d_0, f_0 can be defined by solution of the system of algebraic equations of the 5 order which can be obtained by the equate coefficients of the equation (2) with the same order of 's'.

For the case where $K_{n_e} = 0{,}01$; $\tau = 0{,}15$ sec; $\omega_i = 0{,}2$ rad/sec the frequency response characteristics of the aircraft optimal dynamics is shown on fig 5.

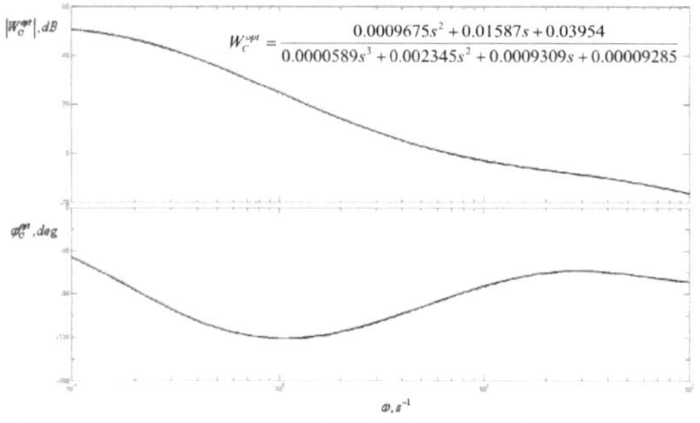

Fig 5. Frequency response of optimal control dynamics

It was offered to include the filter W_f in flight control system for the improvement of the accuracy. Parameters of the filters are selected from the requirement of closeness of controlled element dynamics to the optimal one W_C^{opt}. The controlled element dynamics are the dynamics of aircraft with DLC system providing the mode of motion – $\Delta\alpha = \text{var}, \Delta\theta = 0$, and additional filter. The general equation for this filter is the following $W_f = K_1 \frac{T_1 s + 1}{T_2 s + 1}$. The selection of the filter parameters was carried out by minimization of criteria:

$$\mathrm{I} = \min_{T_1, T_2} \sum_{w_i} \left[\left\| W_C \right| - \left| W_C^{opt} \right\|^2 + \alpha(\varphi_c - \varphi_{c\,opt})^2 \right] \quad (3)$$

where $|W_C|$ and $|W_C^{opt}|$ – amplitude frequency response characteristics of controlled element dynamics and aircraft optimal dynamics, φ_c and $\varphi_{c\,opt}$ – phase frequency response characteristics of the same dynamics, α – weighting coefficient value equal to $\frac{1}{57{,}3}$. The frequencies ω_i are the frequencies of polyharmonic input signal $i(t) = \sum_k A_k cos(\omega_k t)$ widely used in MAI researches [5, 9] for pilot-aircraft system analysis. The values of the filter parameters are given in the table 2 for the case of actuator with ideal dynamics. For the both versions of DLC system these paramiters are the same (K_1=1.46, T_1=0.34, T_2=0.1) and the actuator dynamics does not influence on these values.

Experiments executed on the workstation demonstrated that installation of additional filter W_f allowed to decrease the variance of error in 1,6-1,7 times (fig 6). For the case when the rate limit was taken into account in the actuator dynamic the effectiveness of the additional filter is higher in 1,5 times.

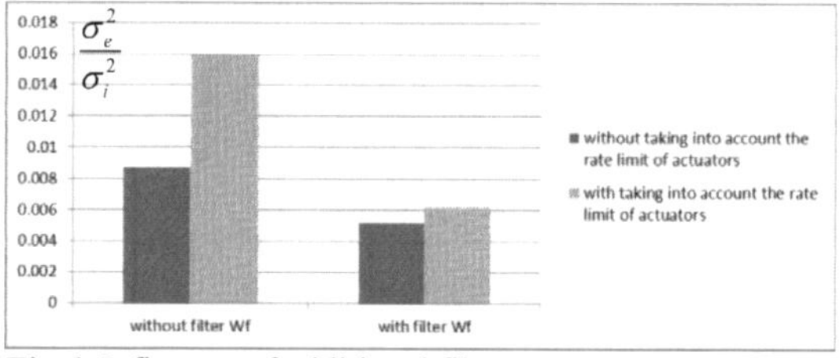

Fig 6. Influence of additional filter

3 The usage of predictive display and its integration with DLC

Two versions of predictive information were considered. The first one ① (fig 7) supposes that the on-board computer generates the predictive vector of the path angle $\gamma_{pr} = \gamma + \dot{\gamma} T/2$.

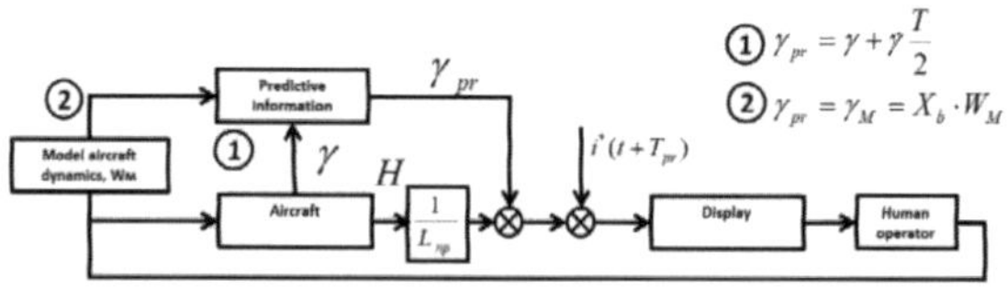

Fig. 7. Pilot-vehicle system with predictive displays

Here the path angle γ is estimated by the inertial system and $\dot{\gamma}$ is the calculated derivative of the path angle. Through the display pilot perceives the predictive aiming angle $\varepsilon_{pr} = \Delta H / L + \gamma_{pr}$, where $L = VT_{pr}$ (T_{pr} – predictive time, V – aircraft velocity) is the distance between the pilot's eye and the surface perpendicular to the 3D corridor generating on the display and limiting the space where the aircraft, has to move.

The other version of predictive information ② (fig 7) supposes that the on-board computer calculates the path angle γ_m according to the mathematical model (W_M, fig 7,2) of the aircraft motion W_M. In that case the pilot perceives the angle $\varepsilon_{pr} = \Delta H/L + \gamma_{pr}$ on the screen of display. Here $\gamma_m = X_e W_M$.

In case when the time delay is accompanied the control process the transfer function of the controlled element dynamics is equal to

$$\frac{\varepsilon}{X_e} = \frac{K_c\left(T_{np}s^2 + 2s + \dfrac{2}{T_{np}}\right)}{2s^2(s^2 + 2\xi\omega s + \omega^2)} e^{-\tau s}.$$

For the second version of predictive display W_M does not include time delay. The frequency response characteristics for these two versions are shown on. fig. 8 for $\tau = 0.18$. It is seen that in case when time delay exist in dynamics the second version allows to suppress its effect.

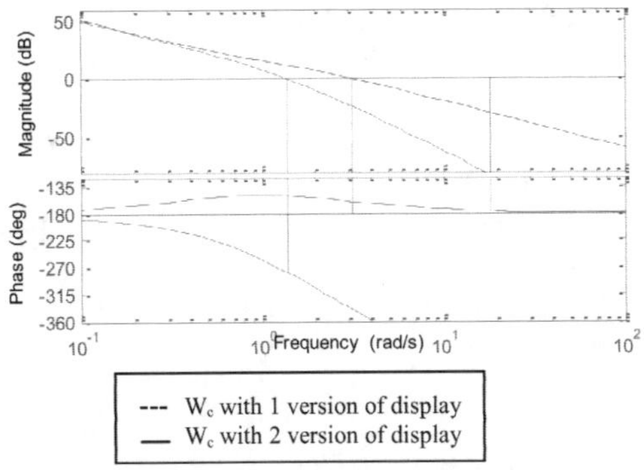

Fig 8. Influence of predictive display on $W_C(jw)$

The determination of predictive display efficiency requires the preliminary definition of predictive time T_{pr} (or $L_{pr} = T_{pr}V$). For that purpose it was developed the procedure based on the modeling of the pilot-vehicle system shown on fig 9.

Here $i(t)$ is the program trajectory and $i(t+\tau)$ is the predictive trajectory projected on display. The optimal T_{pr} (L_{pr}) was defined as the value corresponding to the minimum variance $\sigma^2_{\Delta H}$, where $\Delta H = H - i(t)$. For the determination of $T_{pr_{opt}}$ it was carried out the mathematical modeling of the single loop pilot-vehicle system for each value T_{pr}.

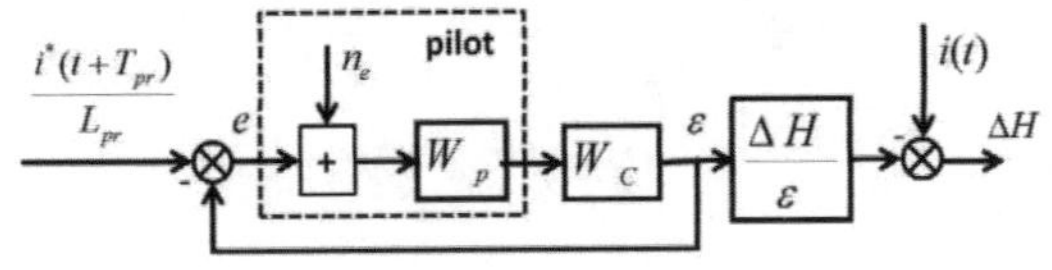

Fig 9. Pilot aircraft system for calculation $\sigma^2_{\Delta_H}$

For that purpose it was used the pilot structural model [5]. The result of such mathematical modeling are the pilot model parameters ($K_p, T_L, \ldots$) used then for calculation of variance of $\sigma^2_{\Delta_H}$ and for definition of its minimum (fig. 10).

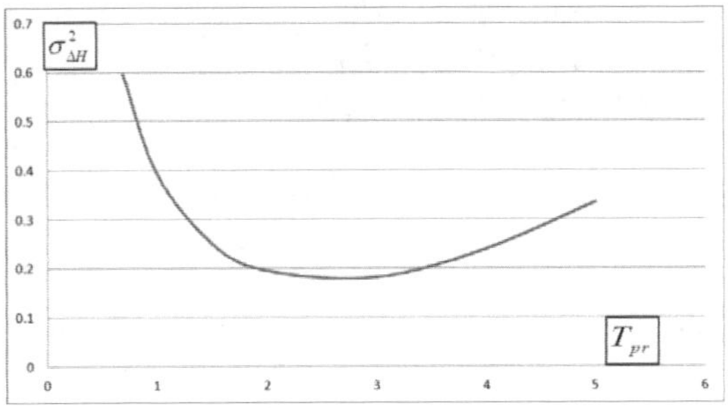

Fig 10. Variance of $\sigma^2_{\Delta H} = f(T_{pr})$

The efficiency of two predictive display versions were studied on the specialized simulator (fig 11). The piloting task was the landing task where the pilot had to track the glide slope and to compensate the error signal ε_{pr}.

Fig 11. Specialized simulator

The results of experiments executed without the disturbances demonstrated that the both versions of predictive display provide approximately the same accuracy in glide slope tracking (fig 12, case $\tau = 0$).

In case when τ was not zero ($\tau = 0.18$) the usage of the second version of display allowed to suppress the effect of controlled element dynamics and to achieve higher accuracy in 1.7 times.

The simultaneous usage of DLC system and predictive display improves the accuracy on about 10-15%. In case when the disturbances were not zero the effect of integration DLC system and predictive display increases up to 40%.

In particular all these results can be explain with help of fig 13. There are given the values of criteria I (3) for the different cases studied in this research. This criteria defines the difference between the amplitudes and phases of the alternative dynamics and optimal frequency response characteristics.

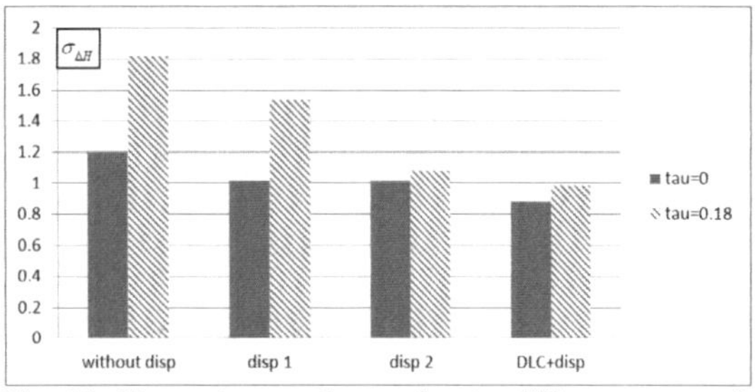

Fig. 12. The variances of error for the different means of augmentation (disp 1 – first version of predictive display, disp 2 – second version of predictive display)

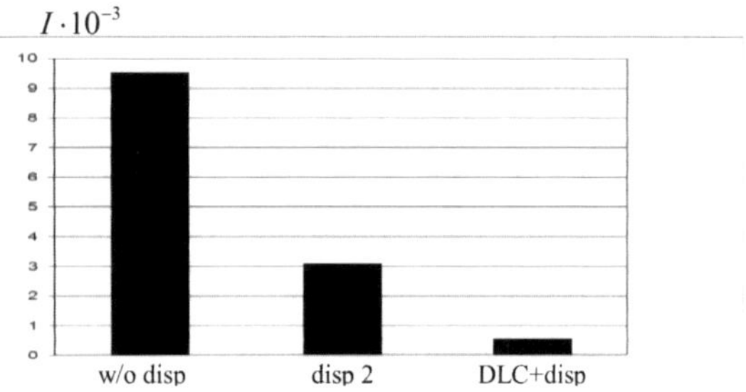

Fig.13. Variances of criteria I (see equation (3))

It is seen that the integration of the DLC system and predictive display provides the highest closeness of the controlled element dynamics to the optimal one.

The results included in the paper are the results of the investigation sponsored by the Ministry of Education and Science (contract № 9.1081.2014/K).

References

1. J.P. Guskov, G.I. Zagainov, *Aircraft Flight Control* (Moscow, Machinostroeniye, 1980)
2. A.V. Efremov, A.V. Ogloblin, *The Ways for Provision of Necessary Flying Qualities. In the book "The Problem of Flight Dynamics of Aircraft Provided by DLC"* (Moscow, MAI, 1981).
3. A.V. Efremov et al., *Journal Russian Aeronautics* **2** (2016)
4. G. Sachs, *Journal of Guidance, Control, and Dynamics*, **23**(3) (2000)
5. A.V. Efremov, M.S. Tjaglick, *The Development of Perspective Displays for Highly Precise Tracking Tasks. In the book "Advances in Aerospace Guidance, Navigation and Control"* (Springer, 2011).
6. A.V. Efremov et al., *Pilot as a Dynamic System.* (Moscow, Machinostroeniye, 1993)
7. J.P. Guskov, *Analysis of Requirements to the Aircraft Handling Qualities Provided by DLC in the Stabilization Task. In the book "The Problem of Flight Dynamics of Aircraft Provided by DLC"*, (Moscow, MAI, 1981)
8. R. Hoh, J. Ashkenas, et al. *Development of Handling Qualities Criteria for Aircraft with Independent Control of Six Degrees of Freedom.* AFWAL-TR-80 (1980).
9. A.V. Efremov et al. Investigation of Pilot Induced Oscillation Tendency and Prediction Criteria Development. WL-TR-96-3109 (Wright Lab USA, 1996).

Airplane Crew Support Implementation Basing on Ellipsoidal Model of the Closed "Pilot-Airplane" Ergatic System

Veniamin N. Evdokimenkov[1], *Roman* Kim[1], *Mikhail N.* Krasilshchikov[1] and *Oleg N.* Korsun[2,*]

[1]Moscow Aviation Institute (National Research University), 125080 Moscow, Russia
[2]FGUP "GosNIIAS", 125319 Moscow, Russia

Abstract. The paper discusses the problem of closed-loop ergatic "pilot - airplane" system monitoring for the purposes of flight safety. The new approach is proposed, based on the confidential ellipsoidal model, which parameters are estimated by flight data processing. The ellipsoidal model of ergatic "pilot - airplane" system enables to detect system emergency operation mode and to identify the source of such emergency mode. The operability of the proposed approach is confirmed by the experimental data processing.

1 Introduction

Analysis of aviation accidents shows, that they are mainly caused by the following factors:
– on-board failures of the airplane and its equipment (22% of aviation accidents);
– pilot's misunderstanding of the current flight situation and resulting pilot's inadequate response (20% of aviation accidents);
– pilot's control errors, that is the manipulating the control levers in unfit mode (18% aviation accidents).

All the above mentioned factors refer to the general ergatic "pilot-airplane" system, that is the closed-loop system which includes pilot, plane, engines, control system, cockpit's indication etc. It is clear that these factors are crucial from the viewpoint of flight safety, so the mathematic model of this ergatic system may be of a great value. The paper proposes a new model for onitoring the state of the closed-loop ergatic system "pilot-airplane" based on the concept of multidimensional confidence ellipsoids. This ellipsoidal model of ergatic system is a intellectual kernel of the suggested approach. This model provides an opportunity to predict flight safety threat. The parameters of the ellipsoidal model of ergatic "pilot-airplane" system are estimated using the flight data.

2 Ellipsoidal model of closed ergatic "pilot-airplane" system and its use for flight safety threat identification

Let us assume, that ergatic system "pilot-airplane" state vector in every time instant t has a block structure $\mathbf{Z}^{\mathrm{T}}(t)=\left(\mathbf{X}^{\mathrm{T}}(t),\mathbf{U}^{\mathrm{T}}(t)\right)$, where vector $\mathbf{X}(t)$ includes parameters of the aircraft motion, and vector $\mathbf{U}(t)$ includes pilot's generated controls such as pitch control stick, throttle lever, pedals, etc. All the components of these vectors are measured and saved by the on-board system of flight data measurement and registration.

Let's consider a typical flight task, for example, the landing of the aircraft. Let us assume, that in the multidimensional space, corresponding to the state vector $\mathbf{Z}^{\mathrm{T}}(t)=\left(\mathbf{X}^{\mathrm{T}}(t),\mathbf{U}^{\mathrm{T}}(t)\right)$, there is a subspace $E(t)$ which principal property is the following:

If in any instant of time t ergatic system state vector $\mathbf{Z}(t)$ belongs to a subspace $E(t)$ the considered flight task may be completed successfully with great probability β, which is close to 1. Let us denote this subspace $E(t)$ as accessibility subspace [1-5].

We can also use the approximation of this accessibility subspace in the multidimensional space in the form of parallelepiped as well as sphere or correlation ellipsoid with a given probability measure [4].

Let us consider the approximation of accessibility subspace $E(t)$ as the correlation ellipsoid $E^{\beta}(t)$ with given probability measure β:

$$E^{\beta}(t)=\{\mathbf{Z}(t):(\mathbf{Z}(t)-\mathbf{m}_Z(t))^T\mathbf{K}_Z^{-1}(\mathbf{Z}(t)-\mathbf{m}_Z(t))\le R_Z(\beta)\},\quad (1)$$

Where $\mathbf{m}_z(t)$ is mathematical expectation of "pilot-airplane" system state vector at instant t for all the flight paths corresponding to successful flight task completion; $\mathbf{K}_Z(t)$ is covariation matrix of "pilot-airplane" system state vector at instant t for all the flight paths corresponding to successful flight task completion; $R_Z(\beta)$ is specific dimension of correlation ellipsoid, which correspond to confidential probability β.

Let us further refer the $E^{\beta}(t)$ (1) as a confidential ellipsoid.

The above defined confidential ellipsoid $E^{\beta}(t)$, approximates the accessibility subspace $E(t)$ and enables to consider statistical relation between airplane flight

* Corresponding author: marmottoduo@yandex.ru

parameters and pilot's control signals. It would make possible to carry out the identification of flight safety threats. Algorithms for estimating confidential ellipsoid $E^{\beta}(t)$ parameters using the flight data are detailed in [5].

The confidential ellipsoid $E^{\beta}(t)$ being formed , it is possible for every instant t of the flight path to estimate the process of piloting from the viewpoint of the flight safety. For this purpose let's assume that measurement errors of vector $\mathbf{Z}(t)$ components are negligible and may be used instead of this vector $\mathbf{Z}(t)$. If this assumption is not valid it's always possible to use Kalman filtering or any other statistical estimators [2]. For systematic errors methods like [6] schould be applied. So, the flight safety analysis of ergatic "pilot-airplane" system is reduced to checking the condition:

$$\mathbf{Z}(t) \in E^{\beta}(t). \tag{2}$$

If for an an arbitrary time instant t flight data meets the condition (2), the ergatic "pilot-airplane" system is operating normally, i.e. any threats to flight safety are absent with guaranteeing probability β . In the opposite, the break of the condition (2) should be treated as an indication of emergency operation mode in the closed-looped "pilot-airplane" contour. So, the condition (2) is integral indicator of "pilot-airplane" mode of operation. However, condition (2) does not enable to identify the cause of emergency situation.

As it was said above, these causes may be divided into three groups:

– on-board failures of the airplane and its equipment;
– pilot's misunderstanding of the current flight situation;
– pilot's control errors, i.e. the insufficient pilot's control activities coordination.

To generate formal criteria for emergency causes identification, it's necessary to consider the formulas (1) and (2) in a detailed way. First we are to note that the correlation ellipsoid (1) and the condition (2) are equivalent to the following inequation:

$$(\mathbf{Z}(t)-\mathbf{m}_Z(t))^T \mathbf{K}_Z^{-1}(\mathbf{Z}(t)-\mathbf{m}_Z(t)) \leq R_Z(\beta). \tag{3}$$

Considering block structure of the vector $\mathbf{Z}(t)$, the correlation matrix $\mathbf{K}_Z(t)$ may also be presented in blocks:

$$\mathbf{K}_Z = \begin{bmatrix} \mathbf{K}_{XX} & \mathbf{K}_{XU} \\ \mathbf{K}_{UX} & \mathbf{K}_{UU} \end{bmatrix}. \tag{4}$$

Let's designate the inverse matrix $\mathbf{K}_Z^{-1}$ as follows:

$$\mathbf{K}_Z^{-1} = \begin{bmatrix} \mathbf{P}_{XX} & \mathbf{P}_{XU} \\ \mathbf{P}_{UX} & \mathbf{P}_{UU} \end{bmatrix}. \tag{5}$$

Using evident relation

$$\mathbf{K}_Z \mathbf{K}_Z^{-1} = \mathbf{E}, \tag{6}$$

where $\mathbf{E}$ is the unit matrix, we can write analytic expressions for elements of inverse matrix $\mathbf{K}_Z^{-1}$:

$$\begin{aligned} &\mathbf{P}_{XX} = (\mathbf{K}_{XX} - \mathbf{K}_{XU}\mathbf{K}_{UU}^{-1}\mathbf{K}_{UX})^{-1}, \\ &\mathbf{P}_{XU} = -\mathbf{K}_{XX}^{-1}\mathbf{K}_{XU}(\mathbf{K}_{UU} - \mathbf{K}_{UX}\mathbf{K}_{XX}^{-1}\mathbf{K}_{XU})^{-1}, \\ &\mathbf{P}_{UU} = (\mathbf{K}_{UU} - \mathbf{K}_{UX}\mathbf{K}_{XX}^{-1}\mathbf{K}_{XU})^{-1}. \end{aligned} \tag{7}$$

The inequation (3) may be represented in the following form:

$$\begin{aligned} &(\mathbf{X}(t)-\mathbf{m}_X(t))^T \mathbf{P}_{XX}(\mathbf{X}(t)-\mathbf{m}_X(t)) + \\ &2(\mathbf{X}(t)-\mathbf{m}_X(t))^T \mathbf{P}_{XU}(\mathbf{U}(t)-\mathbf{m}_U(t)) + \\ &(\mathbf{U}(t)-\mathbf{m}_U(t))^T \mathbf{P}_{UU}(\mathbf{U}(t)-\mathbf{m}_U(t)) \leq R_Z(\beta) \end{aligned} \tag{8}$$

Let's involve the following designation:

$$\begin{aligned} &R_{XX}(t) = (\mathbf{X}(t)-\mathbf{m}_X(t))^T \mathbf{P}_{XX}(\mathbf{X}(t)-\mathbf{m}_X(t)) \\ &R_{XU}(t) = 2(\mathbf{X}(t)-\mathbf{m}_X(t))^T \mathbf{P}_{XU}(\mathbf{U}(t)-\mathbf{m}_U(t)) \\ &R_{UU}(t) = (\mathbf{U}(t)-\mathbf{m}_U(t))^T \mathbf{P}_{UU}(\mathbf{U}(t)-\mathbf{m}_U(t)) \end{aligned} \tag{9}$$

As a result, condition (3) may be presented in the form:

$$R_{XX}(t) + R_{XU}(t) + R_{UU}(t) \leq R_Z(\beta) \tag{10}$$

It's clear that in place of the inequation (10) it's possible to consider the following system of inequations:

$$\begin{aligned} &R_{XX}(t) \leq R_{XX}^{\alpha}(t), \\ &R_{XU}(t) \leq R_{XU}^{\alpha}(t), \\ &R_{UU}(t) \leq R_{UU}^{\alpha}(t), \end{aligned} \tag{11}$$

where parameters $R_{XX}^{\alpha}(t)$, $R_{XU}^{\alpha}(t)$, $R_{UU}^{\alpha}(t)$ correspond to guaranteeing probability $\alpha = \sqrt[4]{\beta}$.

The estimates for $R_{XX}^{\alpha}(t), R_{XU}^{\alpha}(t), R_{UU}^{\alpha}(t)$ are determined by processing the flight data $\mathbf{X}^j(t), \mathbf{U}^j(t), j = 1, \ldots, N$ according to the algorithms presented in [7]. Let's discuss conditions (11) in a detailed way.

Condition

$$R_{XX}(t) = (\mathbf{X}(t)-\mathbf{m}_X(t))^T \mathbf{P}_{XX}(\mathbf{X}(t)-\mathbf{m}_X(t)) \leq R_{XX}^{\alpha}(t) \tag{12}$$

describes for an arbitrary time instant t statistical relation between components of the airplane state vector $\mathbf{X}(t)$. This vector includes flight parameters which depend on the state of the airplane, engines and on-board equipment. If the inequation (12) is not satisfied, the airplane and the on-board facilities are in emergency mode of operation. In other words, the condition (12) is a state indicator for the technical part of the ergatic "pilot-airplane" system.

Condition

$$R_{XU}(t) = 2(\mathbf{X}(t)-\mathbf{m}_X(t))^T \mathbf{P}_{XU}(\mathbf{U}(t)-\mathbf{m}_U(t)) \leq R_{XU}^{\alpha}(t) \tag{13}$$

reflects statistical relation between airplane state parameters $\mathbf{X}(t)$ and control parameters $\mathbf{U}(t)$. This relation depends on the degree of compatibility between the pilot control activities and the current state of the airplane. If the inequality (13) is not satisfied it may be a proof of insufficient pilot's attentiveness as well as of

DOI: 10.1051/matecconf/20179903006

pilot's ignoring some information signals. Moreover, analysis of signals, which break the condition (13) enables to localize information signal ignored by the pilot. In this case a warning should be generated in order to concentrate the pilot's attention.

Inequation

$$R_{UU}(t) = (\mathbf{U}(t) - \mathbf{m}_U(t))^T \mathbf{P}_{UU}(\mathbf{U}(t) - \mathbf{m}_U(t)) \le R^{\alpha}_{UU}(t) \quad (14)$$

reflects statistical relation between various pilot's controls (throttle lever, pitch control stick, etc.) If for an arbitrary time instant t the inequality (14) is not satisfied it may be a proof of the insufficient pilot's control activities coordination. Moreover, analysis of condition (14) enables to identify the clumsy control signals. In this case the generation of a warning is also recommended.

3 Some examples of "pilot-airplane" system state estimation, based on ellipsoidal model

The practical use of the described above ellipsoidal model for identification of flight safety threats requires statistical estimates of the model parameters $\mathbf{m}_X(t)$, $\mathbf{m}_U(t)$, $\mathbf{P}_{XX}(t)$, $\mathbf{P}_{XU}(t)$, $\mathbf{P}_{UU}(t)$, $R^{\alpha}_{XX}(t)$, $R^{\alpha}_{XU}(t)$, $R^{\alpha}_{UU}(t)$. It means that for every piloting task a great amount of flight data is needed. The investigations, carried out by the authors, confirm that the proposed approach is valid when the considerable part of the data is generated in the piloting simulation facilities. Naturally, it's obligatory that the flight data and the simulation correspond to the same type of the airplane, the same piloting task and the same pilot.

As an example let us consider such an important piloting task as the landing. Let us assume that the necessary amount of the flight data is obtained from real flights, simulation or the both. Since the landing path and the airspeed of the aircraft are subject of a strong regulation, there is a stable relation between time t and the distance l from the airplain and the runway center. It means that in expressions (1)-(14) for ellipsoidal model of closed-loup ergatic "pilot-airplane" system it is possible to use distance l instead of time t.

So, let's suppose that the flight data is presented in the form of realizations $\mathbf{X}^j(l), \mathbf{U}^j(l), j = 1, \ldots, N$, which are the result of landing flight mode, obtained by actual flight and simulation. Let's assume that the estimates for $\mathbf{m}_X(l)$, $\mathbf{m}_U(l)$, $\mathbf{P}_{XX}(l)$, $\mathbf{P}_{XU}(l)$, $\mathbf{P}_{UU}(l)$ are calculated, using standard statistical approach for samples $\mathbf{X}^j(l)$, $\mathbf{U}^j(l), j = 1, \ldots, N$. Corresponding algorithms are detailed in [7].

The ellipsoidal model, generated on the basis of $\mathbf{m}_X(l)$, $\mathbf{m}_U(l)$, $\mathbf{P}_{XX}(l)$, $\mathbf{P}_{XU}(l)$, $\mathbf{P}_{UU}(l)$, $R^{\alpha}_{XX}(l), R^{\alpha}_{XU}(l), R^{\alpha}_{UU}(l)$ may be used for identification of flight safety threats.

In order to obtain the detailed identification parameters

$$R_{XU}(l) = 2(\mathbf{X}(l) - \mathbf{m}_X(l))^T \mathbf{P}_{XU}(\mathbf{U}(l)) - \mathbf{m}_U(l))$$
$$R_{UU}(l) = (\mathbf{U}(l) - \mathbf{m}_U(l))^T \mathbf{P}_{UU}(\mathbf{U}(l) - \mathbf{m}_U(l)) \quad (15)$$

are to be calculated, using actual values of $\mathrm{X}(l)$ and $\mathrm{U}(l)$ vectors, where l is distance from center of runway.

The obtained values $R_{XU}(l)$ and $R_{UU}(l)$ for an arbitrary value l should be compared with corresponding values of $R^{\alpha}_{XU}(l), R^{\alpha}_{UU}(l)$. The results are presented on Fig. 1-4. Fig. 1 shows $R_{XU}(l)$ for landing modes, which have been completed according to all landing accuracy requirements. Fig. 2 shows corresponding relations for landing modes, in which the landing accuracy requirements were broken due to insufficient pilot's attentiveness. Fig. 3 presents $R_{UU}(l)$ for landing modes, which have been completed according to all landing accuracy requirements. Fig. 4 gives corresponding relations for landing modes, which have been completed with the break of landing accuracy requirements as the result of insufficient coordination of pilot's control activities.

Analysis shows, that for landing modes, completed according to landing accuracy requirements (Fig.1, 3) at any arbitrary instant of landing the following conditions are met:

$$R_{XU}(l) < R^{\alpha}_{XU}(l),$$
$$R_{UU}(l) < R^{\alpha}_{UU}(l).$$

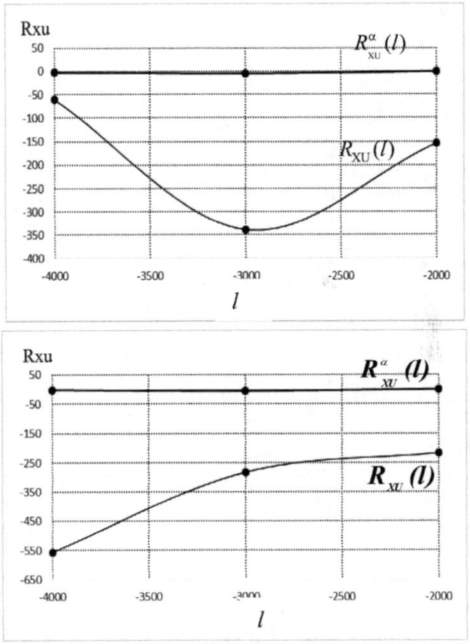

Fig. 1. $R_{XU}(l)$, $R^{\alpha}_{XU}(l)$ for landing modes, satisfying the landing accuracy requirements

In contrast, for landing modes, completed with the break of landing accuracy requirements at some instants of landing process corresponding conditions are not met (Fig.2, 4):

$$R_{XU}(l) > R^{\alpha}_{XU}(l),$$
$$R_{UU}(l) > R^{\alpha}_{UU}(l).$$

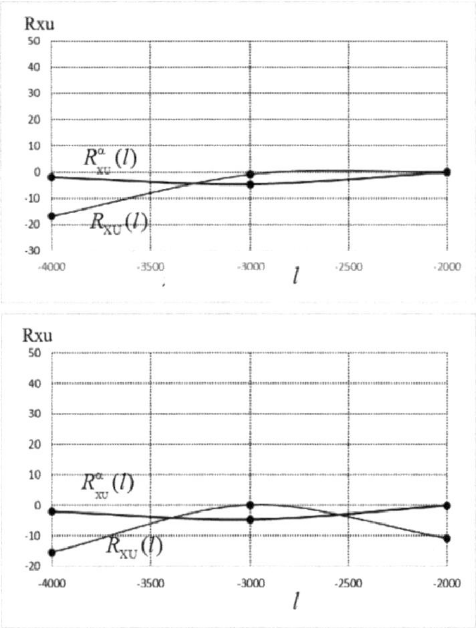

Fig. 2 $R_{XU}(l)$, $R^{a}_{XU}(l)$ for landing modes, breaking landing accuracy requirements

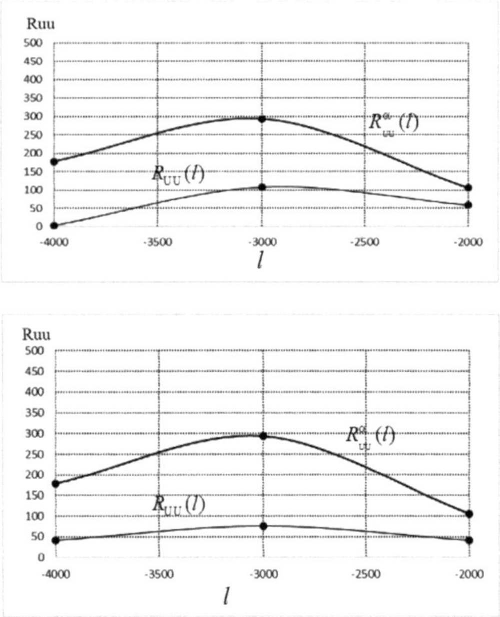

Fig. 3. $R_{UU}(l)$, $R^{\alpha}_{UU}(l)$ relations for landing modes, satisfying the landing accuracy requirements

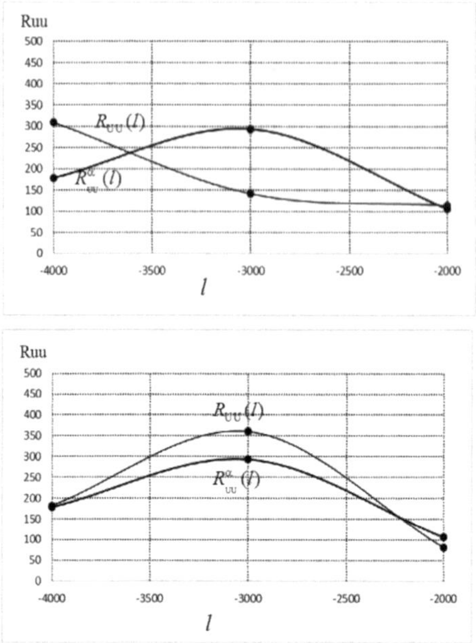

Fig. 4. Relations $R_{UU}(l)$, $R^{\alpha}_{UU}(l)$ relations for landing modes, breaking the landing accuracy requirements

4 Conclusion

Thus, the processing of the experimental results confirm that the proposed ellipsoidal model of ergatic "pilot-airplane" system provides the detection of emergency modes of operation and also the identification of flight safety threats.

References

1. F.L. Chernousko, *Estimation of dynamic system phase state* (Moscow, Nauka, 1988).
2. I.A. Boguslavskiy. *Journal of Computer and Systems Sciences International*, **35**(4), 511-521 (1996)
3. V.V. Malyshev, A.I. Kibzun, *Analysis and synthesis of high accurate verifying vehicle control* (Moscow, Mashinostroenie, 1987)
4. V.V. Malyshev, M.N. Krasilshchikov. V.I. Karlov, *Optimization of observation and control of flying vehicles* (Moscow, Mashinostroenie, 1989)
5. V.N. Evdokimenko., R.V. Kim, M.N. Krasil'shchikov, G.G. Sebryakov, *Journal of Computer and Systems Sciences International*, **54**(4), 609-620 (2015)
6. O.N. Korsun, S.V. Nikolaev, S.G. Pushkov, *Journal of computer and systems sciences international*, **55**(3), 446-457 (2016)
7. M.N. Krasilshchikov, V.N. Evdokimenkov, D.A. Bazlev. *Individually-adapted on-board systems of flying vehicles technical state monitoring and crew support* (Moscow, MAI, 2011).

The New Algorithm for Speech Control in the Cockpit

Aydar Gabdrakhmanov[1,*]

[1] Moscow Institute of Physics and Technology (State University), Moscow, Russia

Abstract. Speech technologies are being developed intensively in the recent years, especially the automatic speech recognition as an additional input method in human interface and technical devices. Most of the known algorithms for speech control have small probability of correct recognition. Widespread methods, like Markov models and neural networks, which require large processing power, allow recognizing the words with a probability of no more than 85–92 %. Such accuracy is not enough to use the voice control on board of a modern aircraft. The article is devoted to a problem of improving the automatic speech recognition's accuracy. A version of word recognition algorithm based on the classical approach is suggested, it includes the comparison with the patterns. In this work to improve the recognition's accuracy a new method of calculating a similarity measurement between the recognizable word and the pattern, which based on z-Fisher transformation, is described. This article also contains an algorithm's modification that takes into account the fixed ratios with the patterns of other words and uses the words adjustment to the pattern with dynamic programming elements. The usage of fixed relations between words provides additional information, which positively affects the recognition. The experimental results of the developed algorithm's approbation on a large amount of speech data are presented.

1 Introduction

In recent years a lot of researches on the improvement of the cockpit interface based on modern audio technologies appeared, for example: control of onboard equipment based on automatic speech recognition [1], the creation of surround sound effect to enhance the information content of a sound indication [2], the analysis in order to ensure the safety characteristics of speech signals [3]. The creation of a cockpit voice interface is complicated by many factors: quality of the existing methods of recognition [1, 4, 5], where the percentage of correctly recognized words rarely reaches 90 - 95%; overload impact [6, 7]; presence of strong acoustic noise [8, 9]; occupational diseases of hearing organs of helicopter aviation crew[10].

2 Formulation of the problem

We propose some algorithms, which can increase the probability of correct recognition, which is critical when using voice control on board the aircraft, where any mistake affects the safety of the flight.

Our algorithms are based on the traditional recognition method [5], which contains the comparison the parametric portraits of words with patterns. Description of the algorithmic implementation given in the article [11]. The conceptual simplicity of this method is useful for evaluating the effectiveness of the proposed new algorithmic solutions. In order to increase the level of recognition we propose the following new approaches:

- A modified measure of closeness between parametric portrait of a recognizable word and patterns. The new measure is to find the middle of the z-Fisher transformation of the correlation coefficients between word's and patter's frames.
- Use of the information about fixed ratios of recognizable words, not only with the appropriate pattern, but also with all other members in the dictionary, which increases the probability of successful recognition.

The initial version of algorithm to the simple case of operator dependent recognition is shown in [12]. This paper proposes a final version of the algorithm and the experimental results in case of operator independent recognition. The effectiveness of the considered algorithms significantly increases when using adaptation of words in the time domain [13], based on dynamic programming approach [4, 14]. Algorithmic implementation is described in detail in [15].

3 Recognition based on comparison with standards

The algorithm is applied classical scheme of finding the maximum value of the measure of closeness between the

* Corresponding author: besha5500@mail.ru

recognizable word and all standards of the dictionary. Let us consider the version used in this study.

Let there be recording of the words in the time domain. For a single word such record has the form

$$x(t_i), i=1,2,\ldots,N_k, \quad (1)$$

$x(t_i)$ – the amplitude of the microphone signal

N_k – number of discrete values of the speech signal.

In our case, the sampling frequency is f=20050 Hz, which corresponds to the sampling interval τ=1/f=1/22050~ 0,05 ms.

Apply to form (1) the spectral-time transformation, which divide the record in time scale into N_t intervals with duration 20…40 ms, on each of which, by using fast Fourier transform algorithm, windows Hannah and averaging over frequencies, compute $N_f = 30..40$ logarithms of signal density values [5, 11]. As a result, we get a parametric portrait of a word in the form of a matrix

$$\mathbf{X} = \{x_{ij}\}, i = 1,2,\ldots,N_f, j = 1,2,\ldots,N_t, \quad (2)$$

where columns correspond N_t quantization intervals in time, and the rows contain values N_f frequency components. Let us apply to the matrix the notation $\{x_{ij}\}$.

In our experiments we use a uniform frequency scale and Mel scale, widespread in the analysis of speech signals [4, 5]:

$$m = 1127\ln(1+\frac{f}{700}),$$

here f – sound frequency in Hz, m – sound frequency in Mel scale.

To create the standard of the word take E implementations of this word and calculate the expected value of E parametric portraits

$$\mathbf{X}_E = \{x_{Eij}\} = \frac{1}{E}\sum_{k=1}^{E}\{x_{ij}^k\}, i = 1,2,\ldots,N_f, j = 1,2,\ldots,N_t \quad (3)$$

The resulting portrait (3) we take as a standard of the word. If it is required to recognize M words, then using formula (3) form M corresponding standards

$$\{x_{Eij}^l\},\ l = 1,2,\ldots,M, \quad (4)$$

where each standard is a matrix of dimension $N_f \times N_t$.

Let it be required to recognize a word that is received in time record (1). Then create the parametric portrait (2) and calculate the measure of proximity (distance) between portrait (2) and each of the standards (4). As the distance we choose, for example, estimate of the correlation coefficient:

$$\hat{r}_{lx} = \frac{\sum_{i=1}^{N_f}\sum_{j=1}^{N_t}(x_{ij}-\hat{m}_x)(x_{ij}^l-\hat{m}_l)}{\sqrt{\sum_{i=1}^{N_f}\sum_{j=1}^{N_t}(x_{ij}-\hat{m}_x)^2 \sum_{i=1}^{N_f}\sum_{j=1}^{N_t}(x_{ij}^l-\hat{m}_l)^2}}, \quad (5)$$

$$\hat{m}_x = \frac{1}{N_f N_t}\sum_{i=1}^{N_f}\sum_{j=1}^{N_t} x_{ij};\quad \hat{m}_l = \frac{1}{N_f N_t}\sum_{i=1}^{N_f}\sum_{j=1}^{N_t} x_{ij}^l, \quad (6)$$

$$l = 1,2,\ldots,M,$$

$\hat{r}_{lx}$ – estimate of the correlation coefficient between recognizable word and standard number l,

$\hat{m}_x$ $\hat{m}_l$ – estimate of the expected value or mean element of the matrix of the portrait of recognizable word, portrait of standard number l,

x_{ij}^l – the element of portrait of standard number l, located at the intersection of row number i and column number j.

The result of recognition is defined by the maximum correlation coefficients among all M standards

$$\hat{r}_{l_{\max}x} = \max_l \hat{r}_{lx}, , \quad (7)$$

$$l = 1,2,\ldots,M$$

$l_{\max}$ – index corresponding to the maximum correlation coefficient and determining the recognition result.

There are other distance measures such as Euclidean distance between the matrices of parametric portraits, but experience shows that they give similar results [1]. Therefore, we will consider this algorithm as a base.

4 Recognition on the basis of comparison with other standards

Considered the traditional recognition scheme is widespread and is to find a standard with a maximum measure of closeness to recognizable word. It reflects the natural process of understanding speech in which a person chooses the most appropriate standard to recognizable word, and does not care about the correlations with other words. This scheme is based on the assumption that a measure of closeness to the word with "its" standard more than with others or "foreign" standards in the dictionary.

When machining in order to increase the probability of correct recognition we offer other schemes of comparison. It is advisable to use the additional information contained in fixed ratios between different words. Note that such a change in the formulation of the problem should not lead to a significant increase in the amount of calculation, as discussed above the detection scheme still requires a comparison with all the standards. Let formulate the new recognition algorithm that uses the results of comparison with all the standards, which are discarded after a maximum location in a traditional scheme.

As part of the basic algorithm (1) - (7) it is advisable to additionally take into account the values of the correlation coefficients between the different words.

Let the training database contains M words with E implementation of each. We form M standards in accordance with (4).

Further, according to (5) for each k word, $k = 1,2,\dots,M$, compute the average by E implementations correlation coefficients with all standards:

$$\bar{r}_{lk} = \frac{1}{E}\sum_{i=1}^{E}\hat{r}_{lk}(i), \qquad (8)$$

$$k = 1,2,\dots,M,\quad l = 1,2,\dots,M,$$

k, l – codes corresponding to the words and standards; $\hat{r}_{lk}(i)$ – estimate of the correlation coefficient between l standard and i implementation of k word; E – the number of realizations of each word in the training database.

Equation (8) defines the matrix of the average coefficients **R** with the size $M \times M$

$$\mathbf{R} = \{\bar{r}_{lk}\},$$

$$k = 1, 2, \dots,M,\quad l = 1, 2, \dots, M$$

in which, for example, the second column consists of average correlation coefficients with all standards and the second word. It turns out that each k word is associated with k column of the matrix $\mathbf{R}$, or vector with dimension M of average correlation coefficients $\mathbf{R}_k = \{\bar{r}_{lk}\}, l = 1,2,\dots,M$.

In recognition of unknown words x calculate from the formula (5) evaluation of the correlation coefficients of its parametric portrait (2) with all M standards, so obtain the vector with dimension M:

$$R_x = \{\hat{r}_{lx}\}, \qquad (9)$$

$$l = 1,2,\dots,M$$

The result of recognition is defined by the minimum norm of subtraction $d_{k_{\min}}$ between the vector (9) and the columns of the matrix $\mathbf{R}$

$$d_{k_{\min}} = \min_k \left\| \{\hat{r}_{lx}\} - \{\bar{r}_{lk}\} \right\|. \qquad (10)$$

Taking the Euclidean norm in the formula (10), we obtain

$$d_{k_{\min}} = \min_k \left\{ \sum_{l=1}^{M} (\hat{r}_{lx} - \bar{r}_{lk})^2 \right\}. \qquad (11)$$

In equations (10) and (11) we search the minimum by all words in the dictionary $k = 1,2,\dots,M$.

5 Modified proximity measure

We create the modified measure of closeness between the recognizable words and standards, which will significantly increase the probability of correct recognition.

Estimates of correlation coefficients have a special asymmetrical distribution [14], so in statistical relation more stable results can be obtained by substituting in formulas (8) - (11) z-Fisher transformation instead of the correlation coefficients, having approximately normal distribution

$$z = \frac{1}{2}\log\left(\frac{1+\hat{r}}{1-\hat{r}}\right), \qquad (12)$$

$\hat{r}$ – evaluation of the correlation coefficient.

Furthermore, let calculate the correlation coefficient not around the hall parametric portrait of the word and its standard as the base in the formula (5), but separately for each time interval (frame) followed by averaging values. Then, for partition to the number of frames N_t calculate for each j frame of recognizable word an assessment of the correlation coefficient with j frame of l standard, going through all the standards $l = 1,2,\dots,M$. In this case, instead of the expression (5) apply the following formula

$$\hat{r}_{lxj} = \frac{\sum_{i=1}^{N_f}(x_{ij} - \hat{m}_x)(x_{ij}^l - \hat{m}_l)}{\sqrt{\sum_{i=1}^{N_f}(x_{ij} - \hat{m}_x)^2 \sum_{i=1}^{N_f}(x_{ij}^l - \hat{m}_l)^2}}, \qquad (13)$$

$$\hat{m}_x = \frac{1}{N_f}\sum_{i=1}^{N_f} x_{ij};\qquad \hat{m}_l = \frac{1}{N_f}\sum_{i=1}^{N_f} x_{ij}^l,$$

$$j = 1,2,\dots,N_t,\quad l = 1,2,\dots,.$$

$\hat{r}_{lxj}$ — evaluation of the correlation coefficient between j frame of recognizable word with j frame of l standard.

Then go to the Fisher z-transform. For this purpose, in the formula (12) substitute estimate of the correlation coefficient (13):

$$z_{lxj} = \frac{1}{2}\log\left(\frac{1+\hat{r}_{lxj}}{1-\hat{r}_{lxj}}\right), \qquad (14)$$

$$j = 1,2,\dots,N_t, l = 1,2,\dots,M.$$

As a final measure of closeness take the average over all intervals N_t z-Fisher transformation (14) of the correlation coefficients (13) between frames of word and the reference:

$$\bar{z}_{lx} = \frac{1}{N_t}\sum_{j=1}^{N_t}\hat{z}_{lxj}, \qquad (15)$$

$$i = 1,2,\dots,N_f, l = 1,2,\dots,M.$$

The proposed two new algorithms 8-11 and 12-15 is applicable in conjunction with the traditional method of pre-adjusting recognizable words in length scale, which reduces the effect of variations in pronunciation of different speakers [10]. The method uses dynamic programming [4, 14, 15].

6 Experimental evaluation of the effectiveness of the developed algorithms

Recognition results of proposed algorithms are tested on the record material of seven different speakers and dictionary with 20 isolated words. Every speaker pronounced 600 words (20 words and 30 realizations of each). Standards formed using similar records of the eighth speaker, so we discuss here speaker independent case of recognition with small learning base composed only one speaker. Recognition results for all speakers and each algorithm are shown in table 1. Where 1 - a basic recognition with a measure of the closeness - the correlation coefficient (6); 2 - a modified proximity measure with averaged z-Fisher transformation (15) of the correlation coefficient; 3 - similar to option 2, but with the implementation of pre-adjusting words in length scale, 4 – the comparison with the standards of "foreign" words, and using modified proximity measure, and pre-adjusting words in length scale.

Table 1. Recognition results (the number of errors in %)

Algorithm	**Speaker**							**The average error rate**
	1	**2**	**3**	**4**	**5**	**6**	**7**	
1	8,2	15,0	12,2	15,0	11,8	8,8	8,0	**11,3**
2	667	9,8	5,5	6,5	9,5	6,0	4,2	**6,9**
3	6,8	7,2	1,0	4,7	5,7	6,0	1,2	**4,6**
4	5,7	5,3	2,0	2,2	4,0	3,5	1,7	**3,5**

Note that each of the listed above algorithmic approach reduces the average error rate. For example, a modified proximity measure reduces the number of errors by 4.4%, pre-adjusting words in length scale together with a modified measure reduce the failure rate by 2.3%, comparing with the "foreign" standards improves results further by 1.1%. Thus, the combined use of three algorithms (modified proximity measure, duration adjustment, comparison with "foreign" standards) allow us to recognize correctly 96.5% of all words. Note that in this experiment we used a small training base composed only one speaker.

7 Conclusions

New approaches to improve the quality of voice recognition are developed:
- a modified measure of closeness between parametric portrait of a recognizable word and patterns; the new measure is to find the middle of the z-Fisher transformation of the correlation coefficients between word's and patter's frames;
- recognition based on a comparison with the "foreign" standards, using fixed relations between different words.

Experimental results show the effectiveness of the proposed algorithms, especially when combined with the method for adjusting the word's length based on dynamic programming.

To the dictionary with 20 isolated words and 30 realizations for each and for seven speakers in a speaker-independent version the combined use of the considered algorithms shows the result up to 96.5% correct recognitions. In this study, the possibility of increasing the quality of recognition, based on the expansion of the training base, deliberately not been used for a better estimation of the effectiveness of the proposed algorithms.

References

1. V. Soshnikov et al., *Trudy GosNIIAS. Seriia: Voprosy avioniki*, **1**, 24–32 (2016)
2. G. Sebryakov et al., *Modelirovanie aviacionnyh sistem: sb. Dokl*, **3**, 452–458 (2011)
3. G. Bondaros et al., *Vestnik komp'iuternykh i informatsionnykh tekhnologii*, **11**, 2–11 (2009)
4. A. Peinado, J. Segura *Speech recognition over digital channels: Robustness and Standards* (New Jersey, John Wiley&Sons, 2006)
5. L. Rabiner, B. Juang, Fundamentals of speech recognition englewood Cliffs (New Jersey, Prentice-Hall International, 1993)
6. A. Ivanov et al., *Vestnik komp'iuternykh i informatsionnykh tekhnologii*, **5**, 3–7 (2012)
7. A. Ivanov et al. *Vestnik komp'yuternyh i informacionnyh tekhnologij*, **6**, 3–7 (2012)
8. V. Chuchupal et al., *Nauka i obrazovanie: nauchnoe izdanie MGTU im. N. Baumana*, **1**, 103–114 (2013)
9. A. Gabdrakhmanov et al., *Nauchnye chteniya po aviacii, posvyashchennye pamyati N. E. ZHukovskogo*, **1**, 330–334 (2013)
10. A. Ivanov et al., *Medicina truda i promyshlennaya ehkologiya*, **11**, 40–45 (2014)
11. A. Gabdrakhmanov et al., *Mekhatronika, avtomatizatsiia, upravlenie*, **9**, 599–604 (2015)
12. O.N. Korsun, A. Gabdrakhmanov, *Nauchnye chteniya po aviacii, posvyashchennye pamyati N. E. ZHukovskogo*, **4**, 331–336 (2016)
13. O.N. Korsun, A. Gabdrakhmanov, *Nauchnye chteniya po aviacii, posvyashchennye pamyati N. E. ZHukovskogo*, **3**, 184–187 (2015)
14. E.S. Wentzel, *Operations research* (DROFA, 2004)
15. O.N. Korsun, A.V. Poliev, *Journal of Computer and Systems Sciences International*, **55**(4), 115–124 (2016)

MATEC Web of Conferences **99**, 03008 (2017) DOI: 10.1051/matecconf/20179903008
CMTAI2016

Linear Isomorphic Regulators

Vladimir Kulabukhov[1,*]

Uspensky Avionica Moscow Research and Production Complex, Moscow, Russia

Abstract: A simple formalized procedure of the synthesis of isomorphic regulators for linear objects "single input - single output". Shown that regulators are isomorphic multiplicative and have the ability to "limit" in terms of quality of transition processes in comparison with additive regulators of the traditional structure.

1 Introduction

The analysis shows [1] that the modern methods of synthesis of regulators are not enough formalized. This applies even to simple linear systems "single input - single output", which are often used in practice. To ensure high quality of transition processes in control loops, such systems need to involve heuristic (artificial) techniques. In order to search for a common solution to this problem, apparatus of modern algebra [2, 3, 4] is frequently used. In this paper [4], based on common algebraic principle of the isomorphy in the theory of systems, the concept of *controllability with accuracy up to some isomorphic model* is introduced. Proved in [4], the theorems have formally allowed to define the structure of regulators for general systems and linear systems. In this article, the method of synthesis of a new type of regulators - isomorphic controllers - is considered based on the results of Article [4], for linear systems of the type "single input - single output". Unlike traditional additive of regulators, isomorphic regulators have multiplicative form. An example of the synthesis of isomorphic controller for linear object is presented. Here are shown the advantages of isomorphic regulators, which include the formal rules of the synthesis and ensuring the maximum achievable quality of transition processes in comparison to traditional regulators.

2 Linear isomorphic regulators

The following system is considered,

$$\dot{X} = A(t)X + B(t)U, X \subset R^n, U \subset R^m, \quad (1)$$

which can be described in Laplace form with zero initial conditions has the form:

$$X(p) = (pI-A)^{-1} BU(p) = \Phi(p)Y(p),$$

where $\Phi(p)=(pI-A(p))^{-1}$ – fundamental (transition) of the system matrix (1) in the Laplace form, I – identity matrix, $Y(p)= B(p)U(p)$- some variable that characterizes the internal state of the system. Matrix $\Phi(p)$ for linear systems is always the opposite, i.e. an isomorphism. The Laplace form of system (1) is converted from the additive to the multiplicative form.

$$\begin{cases} Y(p) = BU(p), \\ X(p) = \Phi(p)Y(p). \end{cases} \quad (2)$$

From [4], that's needed for mappings of matrices from equation (2), commutative diagram has the form and it's shown in Fig. 1.

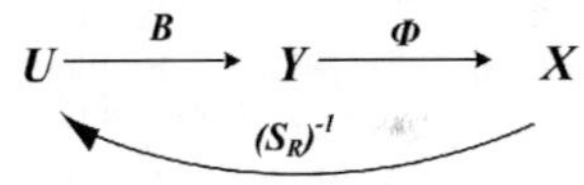

Fig. 1. The composition of mappings to determine the controllability of linear systems.

After Fig. 1, for a linear object (1) in the multiplicative form (2) isomorphic with accuracy up to isomorphism Φ regulator S_R^{-1} will have the form:

$$S_R^{-1} = B^{-1}\Phi^{-1}(p) = B^{-1}(pI-A(p)). \quad (3)$$

Those regulators, unlike known additive regulators that use the addition-subtraction operation will be called isomorphic multiplicative regulators. Conditions of controllability of object (2) and the rules for determining the inverse matrices S_R^{-1} and B^{-1} with accuracy up to isomorphism Φ are defined in [4]. It is shown [4] that for a controlled object within the commutative diagram, there are always inverse matrix S_R^{-1} and B^{-1}, even if the original matrices are rectangular and out of irreversible chart. Suppose that the system is controllable. Require that the actual value of the vector X tracked a desired value $X_{жс}$ in accordance with the movement of the reference model with a transfer function $W_{эм}$. Those problems of synthesis of the prototype have universal character [2]. Reference model is selected, for example, taking into account the quality of engineering criteria.

* Corresponding author: kws0704@mail.ru

Considering various forms of reference model, the equation (1) of the control system take the form:

$$\dot{X} = AX + BU,\ X_0 \quad (4)$$

$$\dot{X}_з = CX_з + DX_{жс},\ X_{з0} = X_0, \quad (5)$$

$$X = IX_з = IX_{жс}, \quad (6)$$

where $X_з$ - "instantaneous" set value, which must exactly match the output X. Reference model equations (5) and (6) correspond to the transfer functions $W_{эм}=(pI-C)^{-1}D$ и $W_{эм}=I$, where I - identity matrix.

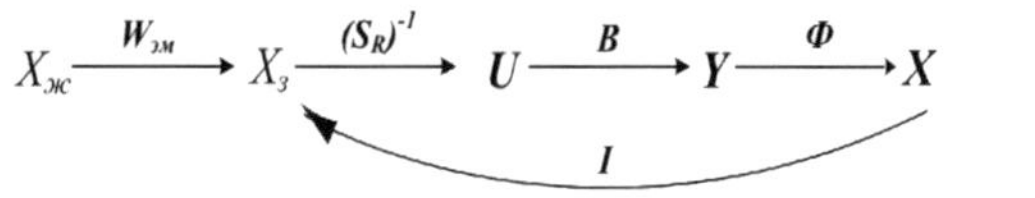

Fig. 2. The composition of mappings for tracking the reference model.

In Fig. 2 it is presented a commutative diagram, obtained from Fig. 1, taking into account (4) - (6). For Fig. 2, equation (3) takes the form;

$$S_R^{-1} = B^{-1}(p)\Phi^{-1}(p)I = B^{-1}(p)\Phi^{-1}(p) = B^{-1}(p)(pI-A(p)),$$
$$S_R = \Phi(p)B(p), \quad (7)$$

where S_R^{-1} and B^{-1} must satisfy conditions [4]:

$$I_x^{right} = S_R S_R^{-1},\ I_x^{left} = \Phi\Phi^{-1},\ I_x^{left} = I_x^{right} = I,$$
$$I_U^{left} = S_R^{-1}S_R,\ I_U^{right} = B^{-1}B,\ I_U^{left} = I_U^{right}. \quad (8)$$

Here I_x^{left} , I_x^{right} - left and right unit on the set of $X_з$=X, I_U^{left} and I_U^{right} - left and right units to the U. In this case, unit on $X_з$ is equal to the identity matrix I. In general, when m$\geq$n I_U^{left} and I_U^{right} can be non-unit matrices in the usual sense, but such that:

$$I_U^{left} = I_U^{right} = I_U = (I_U)^{-1}.$$

Equations (7) and (8) allow to determine an isomorphic regulator S_R^{-1}. If the model (4) exactly matches the object, the control loop with a regulator S_R^{-1} as feedback (FB) which is shown in Fig. 2, acquires a "linear" form (see. Figure 3) without the "cycles" (without visible FB).

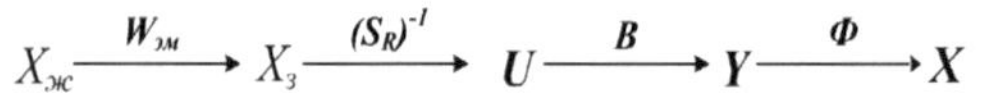

Fig. 3. "Linear" loop with the isomorphic multiplicative regulator.

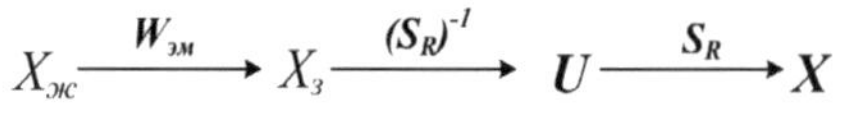

Fig. 4. "linear" loop with maximum achievable control quality.

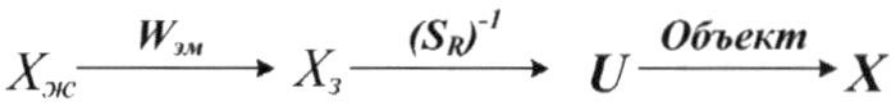

Fig. 5. "Linear" object control loop

It follows from (7) and Fig. 3, the circuit considering equation S_R= $\Phi(p)B(p)$ is converted to the form which is shown in Fig. 4. This "linear" loop realizes condition $S_R S_R^{-1}$=I of accurate tracking X=$X_з$ X=$X_з$ at any time. "Linear" loop does not mean that there is no FB from X to U. The presence and structure of FB "encrypted" in the regulator S_R^{-1}. Unlike traditional additive regulators summators in control loops, a regulator SR-1 is multiplicative; such regulator requires only multiplication of matrices present in the "linear" loop. As can be seen from Fig. 4, object movement with the right synthesized isomorphic to control completely and with absolute accuracy is determined by the movement of the reference model with the transfer function $W_{эм}$. In this sense, the isomorphic regulator is limiting by the quality of transition processes. Since every object has a dynamic and limited power controls, as the preferred option should be considered reference model (5) from $W_{эм}=(pI-C)^{-1}D$. Thus, $X_{жс}$ may comprise only a few components of X. On the other components of X and all components of U may be restricted. Let's consider the case without restrictions.

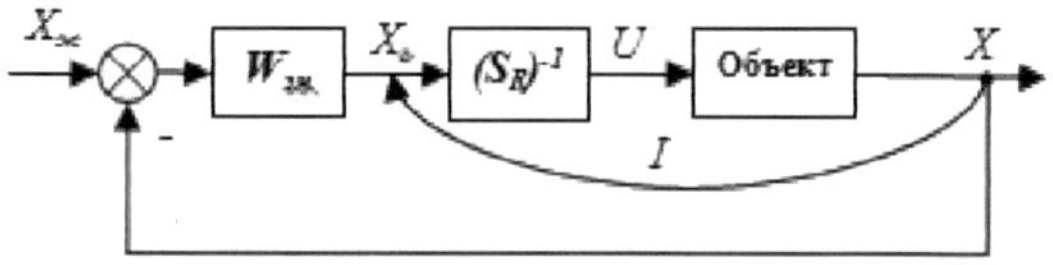

Fig. 6. The robust control loop.

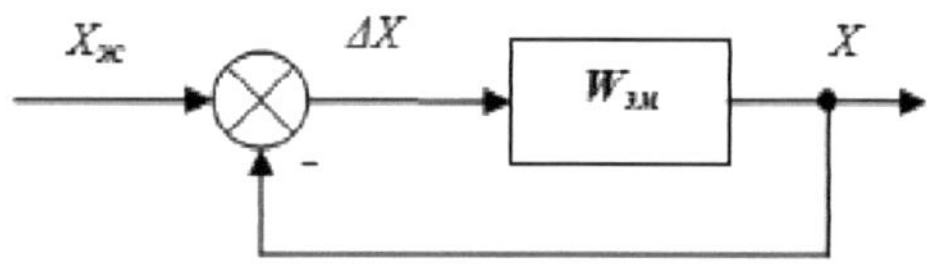

Fig. 7. The robust control loop model with adequate object.

If the objects are controlled and it's obtained "linear" loop shown in Fig. 3-5, it is necessary to check out through modeling it on the robustness, by asking variations of parameters of matrixes Φ and B. If the result is satisfactory, the loop can be used for control. If the object is significantly different from the model (4) used in the synthesis of the regulator, the "linear" loop can be insufficiently robust. Then it is necessary to cover it with additional FB, as shown in Fig. 6. For the adequate model object (4) , transfer function "control object" from $X_з$ of X close to the unit $W_{PO} \approx I$ and a closed loop of $X_{жс}$ to X (see. Fig. 6) will be as shown in Fig. 7, with the transfer function $W= W_{эм}\ (I+ W_{эм})^{-1}$. That is, the characteristics of a robust control loop, as before, will be determined only by the structure of reference model $W_{эм}=(pI-C)^{-1}D$. Next, let's consider an object with one input and one output. In the scalar case, at $W_{ЭМ} = k_м(T_м p+1)^{-1}$, the transfer function of the closed loop will be obtained.

$$W = \frac{k_м}{T_м p + 1 + k_м} = \frac{k_м}{(1+k_м)\left[\frac{T_м}{(1+k_м)}p+1\right]} \approx \frac{1}{T_м^* p+1} \quad (9)$$

Where, $T^*_{\text{м}} = \dfrac{T_{\text{м}}}{(1+k_{\text{м}})}$, which provides astatism on controlled coordinates at sufficiently large $k_{\text{м}}$. At the same time $T^*_{\text{м}}$ is $(1+k_{\text{м}})$ times less than $T_{\text{м}}$ and the transition process in the robust loop in comparison with the "linear" loop is sharply reduced, which could adversely affect the stability and cost controls. To ensure the prior time of transition process in robust loop simultaneously with increment $k_{\text{м}}$, it's necessary to increase $(1+k_{\text{м}})$ times and time constant $T_{\text{м}}$ to $W_{ЭМ}$ times, which has the form $W_{ЭМ} = k_{\text{м}}(T_{\text{м}}(1+k_{\text{м}})p+1)^{-1}$, where $T_{\text{м}}$ - the prior time constant (before the introduction of additional feedback and increase). Then, the transfer function of the closed loop shown in Fig. 7, takes the form:

$$W = \frac{k_{\text{м}}}{T_{\text{м}}(1+k_{\text{м}})p+1+k_{\text{м}}} = \frac{k_{\text{м}}}{(1+k_{\text{м}})(T_{\text{м}}p+1)} \approx \frac{1}{T_{\text{м}}p+1} \quad (10)$$

and at large $k_{\text{м}}$ will practically match exactly the transfer function of the original loop without additional FB.

3 Example

Let's consider the specific view of matrices *A* and *B* in the object model (4). Let

$$A=\begin{bmatrix} a_{11} & a_{12} \\ a_{21} & a_{22} \end{bmatrix} \text{ and } B=\begin{bmatrix} b_{11} & 0 \\ 0 & 0 \end{bmatrix}. \quad (11)$$

Equations (4) and (11) can be described in the scalar form as follow:

$$\begin{cases} \dot{x}_1 = a_{11}x_1 + a_{12}x_2 + b_{11}u_1, \\ \dot{x}_2 = a_{21}x_1 + a_{22}x_2. \end{cases} \quad (12)$$

By considering $x_2 = a_{21}(p-a_{22})^{-1}x_1 = ax_1$, where $a = a_{21}(p-a_{22})^{-1}$, and equation (12) could be transformed in to the form:

$$\begin{cases} x_1 = bu_1, \\ x_2 = ax_1, \end{cases} \quad (13)$$

where $b = \dfrac{b_{11}}{p-a_{11}-a_{12}a}$.

For (4), by considering (11) in the form (13) diagram is shown in Fig. 3. In this case, *B=b, Φ=a.* The transfer function of control object W_O has the form:

$$W_O = s_r = ab = \frac{b_{11}a_{21}}{p^2-(a_{11}+a_{22})p+(a_{11}a_{22}-a_{12}a_{21})} \quad (14)$$

If the transfer function of object is known, then it is possible to obtain "linear" control loop from Fig (5) as shown in Fig (8).

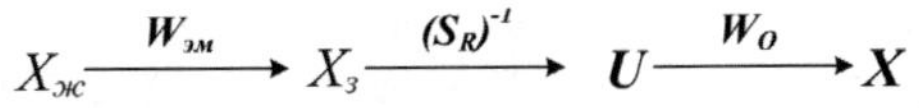

Fig. 8. "linear" loop for the object (13).

From Fig. 8, it implies in the scalar case, for the object of the "single input - single output" regulator definition reduces to the handling of the transfer function of the object, that is,

$$S_R^{-1} = W_O^{-1}. \quad (15)$$

Let the elements of the matrices (11) have the following values:

$$A=\begin{bmatrix} -4 & -1{,}14 \\ 1 & -0{,}62 \end{bmatrix} \text{ and } B=\begin{bmatrix} -7{,}5 & 0 \\ 0 & 0 \end{bmatrix}. \quad (16)$$

Simulation of an ideal loop (see in Figure 8) with the regulator (15) and the type of reference model $W_{ЭМ} = k(Tp+1)^{-1}$ at *k=1* and *T=0,4*, it showed that x_2 tracks accurately given reference model movement. At inaccurate knowledge of elements a_{ij} in (16), in the "linear" loop, working off error x_2 of error order of work a_{ij} appears. Fig (9) shows the transition process (upper graph) at a_{21}=1.25 (+25%). Short loop of additional operating circuit according to Fig. 6-7 gives the loop robustness property. Fig (9) (lower graphic) shows that in the loop with the additional FB at the increased gain factor to a value of k = 99 with a simultaneous increased T in 1 times, according to (10) even at the parameter a21 is increased to 25%, the quality of transition process remains high. Similar results were obtained at varying other matrix elements (16).

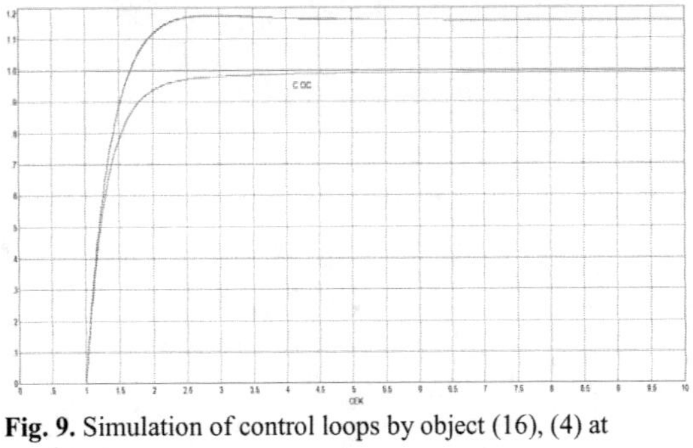

Fig. 9. Simulation of control loops by object (16), (4) at varying a_{ij}.

The simulation showed that the "linear" circuit has suitable robustness and for circuit with additional feedback, range of variation of matrix elements (16), while maintaining the quality of transients can be tens percent. Thus, loop of astatism can be controlled by changing gain factor *k*.

Conclusion

Synthesis rules of isomorphic regulators on the basis of the theory [4] are examined. It is shown that the isomorphic regulators are multiplicative, and the procedure of their synthesis for the object with one input and one output has been formalized as much as possible: synthesis of regulator is reduced to handling of the transfer function of the object. In this case, the required quality of transition processes is provided by selecting of the reference model. In this sense, isomorphic regulators have the property of "limit" on the level of achievable quality of transition processes in comparison with known regulators. An example of the synthesis and simulation of isomorphic regulator showed this dignity.

References

1. V. Bulgakov, V. Kulabuhov. *Instruments (Pribory)*, **1** (151), 39-44 (2013)
2. V. Bukov, *Attachment systems. Analytical approach to the analysis and synthesis of matrix systems*, (2006)
3. D. Kim, *Mechatronics, Automation, Control*, **1**, 9-15, (2011)
4. V. Kulabuhov, *XII Russian Conference on Control Problems VSPU-2014*, pp. 438-448, (2014)

MATEC Web of Conferences **99**, 03009 (2017) DOI: 10.1051/matecconf/20179903009
CMTAI2016

The Use of Uavs Groups for the Monitoring of Terrestrial Infrastructure Objects

German G. Sebryakov[1, *] and *Viliam V.* Insarov[1]

[1]FGUP "GosNIIAS", Moscow, Russia

Abstract. The paper discusses the basic principles of UAF application to the task of monitoring the terrestrial areas. The emphasis is made on the problem of controlling the UAVs groups and the distribution of functions between the UAVs in the group and the on-land control centres.

1 Introduction

Recent developments in the field of intelligent systems made it possible to enlarge UAVs application area. Among those developments are: improved capacity for autonomous flight and flight in groups; reduction of high-precision sensors in size and power consumption, which made it possible to use them on board; improved means of communication between UAVs and between the UAV and base station; etc.

Of particular significance in this context is the application of drones for performing complex tasks, such as traffic surveillance; condition monitoring of oil and gas pipelines or power lines, detection and monitoring of ice jams; data collection on chemical or radioactive contamination of air and terrain surface; engineer reconnaissance missions during or after floods, earthquakes or other natural disasters; as a source of visual information during peacekeeping operations, or evacuation operations; forest fires detection, as information support for firefighting teams, etc.

Special equipment installed on board of the UAV, enables it to conduct operations in the areas of terrain, otherwise inaccessible for humans. For example, highly sensitive radars are able to register movement under piles of debris or avalanches; sensors that register the presence of hazardous substances in the air allow to rapidly define the boundaries of contaminated zone; infrared cameras allow to use the optical recognition of objects in the dark.

It is obvious that for most applications listed above maximum efficiency can be achieved through the use of groups of UAVs as opposed to the use of a single aircraft[1-4].

One more application area of UAV groups should be mentioned here apart from the ones mentioned above is the implementation of so-called task-oriented control architecture. In this case operators interact with a general mission and payload manager, instead of having individually tasking each vehicle. The general manager coordinates the set of tasks for a group of UAVs which typically communicate with one another in a decentralized mode. In a system like that every vechicle in the group can perform it's own task thus improving group's overal situation efficiency and information processing rates.

Group of UAVs, when organized in a form of a spatially distributed system can solve the problem of passive detection of radio-emitting and, above all, moving targets with rather high accuracy [5].

It is believed that a group of drones can be effectively used in the electronic warfare in order to suppress communication channels of the opposing side.

Of course, the methods of organizing the group applications of UAVs in each one of the considered situations have their own particularities. Nevertheless, it is of interest to us to consider some of the general points, common to all the implementations of this approach, mainly because the the most part of tasks described above, have the similar sets of elementary actions to be performed in order to accomplish the task. So, the decomposition into similar subtasks is possible.

2 Formation of a group of UAVs

In order to perform tasks that involve cooperative UAV actions following problems should be solved :

1. Organization of the group of agents into a network-like system with UAVs of the same type or subgroups of UAVs of the same type or subgroups containing different types of UAVs and manned aircrafts.

2. Determining the optimal number of agents in the group as a solution of the optimization problem. It is to be assumed that the time intervals between flights and flight trajectories constraints are known and the flow of applications may be considered as a random process.

3. Organization of UAVs group takeoff, gathering agents into cruise formation, maintaining the relative position of the agents within the formation during the cruise flight, return to the base, landing of the agents, etc.

* Corresponding author: sebr@gosniias.ru

4. Route planning and scheduling with account to it's physical feasibility and global requirements applied to the performance; initial decomposition of tasks into a set of logically related subtasks and further distribution of subtasks between the agents of the group.

5. Creating a software platform for the implementation of mission-planning and task payload management capable of coping with predefined set of subtasks, equiped with a set of protocols for ingroup communication and means to add new functions if needed.

6. Establishing effective communication between group agents and mission control center.

7. Making use of modern multichannel and multispectral on board computer vision systems.

The problems listed above do not include the problems of UAV type selection or selection of it's power unit, on board equipment complex, and so on. The report deliberately focuses on the problems inherent in the use of the group of UAVs in dealing with land surface monitoring tasks.

3 Some notes

Let us discuss some of the subjects mentioned in previous sections of this report, and let us take a cooperative UAV surveillance task as an example. The observed object is considered to be immobile and all it's characteristics are known. If the software platform is implemented according to modular approach [1-3], then all the characteristics of the object and the search area should be provided to it as inputs in order to formulate the search task. This task is then loaded to planning modules of all the group agents. As a result those modules produce the initial mission plan and a list of individual flight plans for all the group members, the plan and the list are then dispatched to the network configuration manager.

During the mission some events may happen, such as the introduction of a new UAV into the group, technical failures, changes in the mission objectives e.t.c. which are not foreseen in the initial plan. Those events should be taken into account by the system; it should adapt and rebuild flight plans accordingly. For the numerical evaluation of the effectiveness of subtasks distribution, we introduce the concept of functional criteria. A particular set of criteria is determined by the nature of the task and can be further adjusted without affecting the general applicability of the planning approach.

For example, let's consider two criteria that allow UAVs at the same time not to gather too densely in one area and not to exceed limits of wireless communication.

Lпi – criterion that characterizes the coverage of the given area by surveillance group agents, it depends on the evaluation of the observability of the given area by the i-th UAV in the group.

Lci – criterion that characterizes the quality of connection between the i-th UAV and other UAVs in group.

In this case, the total efficiency of the system at any step is determined by the sum of efficiencies of all N UAVs in the group, according to the formula

$$L_g = k_p \sum L_{pi} + k_c \sum L_{ci}, \qquad (1)$$

where $i = 1, 2\ldots, N$; N – number of UAVs in the group; L_g – group performance criterion; k_p, k_c – weight coefficients ($k_p + k_c = 1$).

The higher the total index L_g, the better the quality of the solution according to this criteria. Since control tasks handling is decentralized, it possible that the same subtask will be appointed to several drones. In order to avoid such situations a protocol for the resolution of conflicts should be presented.

Naturally, the appropriate data link digital technologies should be used toprovide the necessary information support.

The use of multi-channel systems for solving problems of detection and recognition of objects and complex scenes has significant advantages over single-channel versions. It provides:

– better reliability of the information about geometrical parameteres and brightness levels of observed objects;
– detection of camouflaged objects, as it is much more difficult to make an object invisible in different spectral ranges;
– noise immunity, because it is difficult to maintain jamming simultaneously in different spectral ranges;
– performance in presence of shielding objects or the noisy backgrounds;
– recognition of objects in poor weather conditions because sensors with different spectral ranges have different sensitivity in same weather conditions.

4 The prospects

The prolems to be solved in the nearest future are the following:

1. Determining the methods of forming a group of UAVs as a integral and stable system with distributed structure.

2. The choice of the optimal degree of UAV autonomy in the group. The degree of operator participation is to be determined by alternative logics such as "operator in the control loop" or "operator above the control loop with the right to make the final decision".

3.The development of a high-performance software for the UAV group management and self-organization tasks.

4. Analysis of the information and find the right job in the structure, providing a virtual exchange of information via wireless communication lines in order to maintain the structure and formation a team and replenishing its lost participants and etc.

5. Development of standards regulating the interaction of the operator with a single unmanned platform or with the group of UAVs at all degrees of interoperability.

6. Development of technique for training the highly skilled operators capable to provide management, navigation and targeting UAV in a group.

7. The development of on-board systems of high-precision detection, based on the multi-channel optical-electronic and radar systems for small-size UAVs.

8. Development of an autonomous system of evading potential threats and ensuring the safety of the UAV in flight in relation to the other UAVs and piloted aircrafts, and to the ground obstacles, natural or artifitial .

10. Development of models for simulation the functioning single UAV and UAVs groups.

References

1. P.O. Skobelev, *Mehatronika, avtomatizatciia, upravlenie*, **12**, 33 (2011)
2. Yu.I. Buryak, V.V. Insarov, *Journal of Computer and Systems Sciences International*, **45**(2), 245-262 (2006)
3. N.V. Kim, I.G. Kry`lov, *Trudy` MAI*, **62** (2012) https://www.mai.ru/science/trudy/published.php?ID =35507
4. D.S. Budaev, G.Iu. Voshchuk, N.A. Gusev et al., Coglasovannoe upravlenie gruppoi` BPLA s primeneniem mul`tiagentny`kh tekhnologii` i setecentricheskogo podhoda. *E`lektronny`i` resurs* (2015) www.conf58.mipt.ru
5. V.Y. Bulychev, Y.G. Bulychev, S.S. Ivakina et al., *Journal of Computer and Systems Sciences International*, **54**(4), 593-608 (2015)

 DOI: 10.1051/matecconf/20179903010

Development and Updating of 3D GIS by Using Aerial Photography Systems

Kirill Zhigalov[1, 2, *], *Denis* Efremov[2], *Gregory* Bubnov[2, 3] and *Evgeniy* Pluzhnik[2]

[1]V.A. Trapeznikov institute of Control Science of Russian Academy of Sciences, 117997 Moscow, Russia
[2]Moscow Technological Institute, 119334 Moscow, Russia
[3]Moscow Institute of Physics and Technology, 141701 Dolgoprudny, Moscow region, Russia

Abstract. In the article the main opportunities for engineers at the GIS creation/updating of systems by means of multicopter aerial photography systems are described. Usage of the methods described in the article will allow increasing the speed and informational content of the GIS construction/updating at rather small labor costs. Besides, the described automation of the processes of filming from the unmanned aerial vehicle will allow to supplement three-dimensional models of objects in GIS with earlier inaccessible data.

1 Introduction

Now geographic information systems are widespread in all fields of personal activity. The main requirements for them are efficiency of updating and accuracy of the data containing in them. The basis of all geographic information systems are different cartographic materials (aerial photography, maps, plans and other) describing the area. The area constantly changes as due to activities of the person (development of pits, deforestation, any construction), and due to natural processes (growth of trees, degradation of the rivers and other) [1–3].

In connection with the aforesaid, expeditious updating of GIS is rather an important task.

It should be noted that 3D GIS also includes 3D models of the area (columns, structures, bridges and other). These models, in case of need are also a subject for updating (see Figure 1).

Fig. 1. An example of the 3D screen of GIS

2 Standard ways of the GIS construction/updating

Updating of these 3D GIS-systems requires obtaining coordinates of all changes on the area in any kind and pictures of the objects of the area taken from the different parties at right angle to the copter [4].

Nowadays for the purpose of data acquisition for the GIS updating 2 main methods are used:
- data acquisition in the manual way (a round of all the territory manually with special devices);
- data acquisition by means of earth remote-sensing instruments, generally they are laser scanners and cameras installed on different carriers (satellites, planes, cars).

The first way is basically used when updating small territory of GIS (up to several hectares).

The second method is applicable in case of big areas of a covering of GIS (coal pits, the woods, roads, lines of the power line, cities and so on).

However both of the ways are widely used, they have a number of disadvantages.

While using a manual way of updating GIS, people can only fix the position of objects on the area without a possibility of addition in GIS pictures of a surface, which allows manufacturing only plans/maps.

In case of manufacturing models of difficult objects (see Figure 2) a person can't see some parts (i.e. upper ones), even using difficult devices, such as, the 3D laser scanner.

The last makes an influence for quality of models and their accuracy.

In case of data collection by remote-sensing instruments we can make rather accurate images of the area (aerial photography, see Figure 3) and coordinates of points of a surface (when using 3D laser scanners).

Nevertheless, for production of 3D models of the objects it is necessary to use the pictures made from different foreshortenings and not just from above (see Figure 4.).

* Corresponding author: kshakalov@mail.ru

DOI: 10.1051/matecconf/20179903010

Fig. 2. An example of a difficult object

Besides, receipt of such data is rather costly from the financial point of view.

Fig. 3. An example of the result of aerial photography

Fig. 4. An example of an image of one object from different foreshortenings

3. Possible improvements connected with the application of copter in aerial photography

It is possible to integrate the methods described above in one universal method with the use of modern quad- and hexacopters, with dynamically suspended cameras (see Figure 5). One can use remotely operated small helicopters for this purpose, but the vibration on one motor is rather big for this purpose, and that may affect the quality of final images. In this connection it is recommended to use quad- and hexacopter based systems [5].

Such suspension allows changing position of the camera remotely on a copter system from parallel to the horizon (for shootings of objects of a terrain, see Figure 6) to perpendicular (for shooting of a surface, see Figure 7).

Modern copters allow moving away from the operator on the distance up to 7 kilometres and on the height up to 5000 meters, which allows covering the area of photography up to 140 square kilometres with one copter.

This configuration allows taking pictures of separate objects located at the distance up to 6 kilometers from the operator from all necessary foreshortenings that leads to obtaining necessary information about remote objects quickly.

Fig. 5. An example of a copter with a dynamic suspension of a camera

Fig. 6. An example of the image of an object for modeling captured by a copter

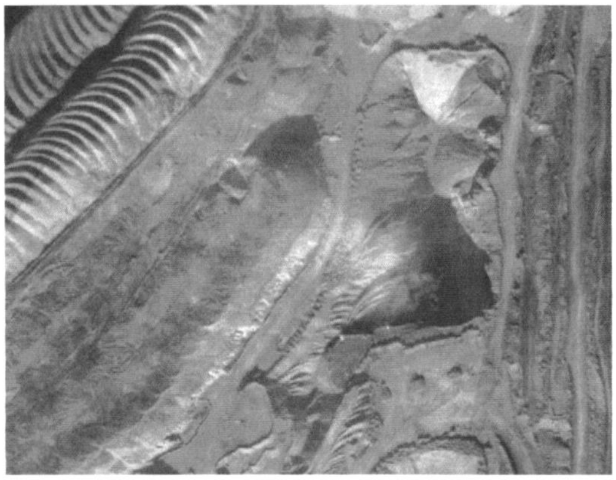

Fig. 7. An example of the image of a surface captured a copter

Nevertheless, there are several disadvantages of data acquisition by means of copter:
- rather big weight of a dynamic suspension with the stabilizer and the camera reduces maneuverability and time of the device usage;
- rather small height of work (in comparison with satellites and the air planes) that leads to a rather large number of pictures of a surface, that causes difficulties at their storage and processing;
- helplessness of modern civil copters in front of the systems of muffling of a signal of management.

The first two difficulties are solved by using modern accumulators with powerful motors, nevertheless they do not strongly influence the final result.

The last difficulty is solved on some copters with the help of automatic return systems and sensors of provision of surrounding objects.

Nevertheless, it is possible to use systems of automated management with the help of integration of digital on-board computer maps and laying of the automated routes of a returning.

As a test attempt of such decision GIS suek3d with a prepared digital district map in advance has been used. The copter several times passed along in advance set flight line (see Figure 8).

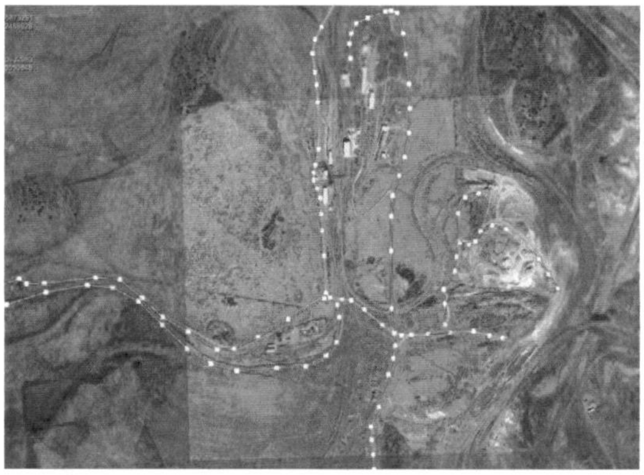

Fig. 8. An example of a modelling flight line in GIS

The test realization on the copter consisted of a control system on the basis of a phone with the MS Windows 10 operating system and a simple code, written on C ++. The data have been transferred to the control system by means of the table with X, Y, Z coordinates for each point of a flight line.

It should be noted, that the system reduced the payload by 200 grams and the flight time by 5 minutes. In such realization it is possible to use less qualified personnel for starts.

4. Conclusion

Updating of 3D GIS data is an important and universal task. Three-dimensional GIS data differ from usual ones by a large number of models of objects and the equipment. Modern ways of data acquisition for the GIS updating are well adapted for updating of standard GIS-systems, nevertheless they have a number of the disadvantages complicating their use in three-dimensional GIS. Copters with small additions in a design are suitable for the new generation GIS updating. The offered option of the usage of a dynamic suspension significantly expands the possibilities of a usual copter and allows to use it as an absolutely new universal platform for data collection for the GIS updating and to accelerate this process. It is especially important in the countries where data collection is limited by climatic conditions.

References

1. A. Bhardwaja, L. Samb, Akankshad, F.J. Martín-Torresa, R. Kumarb, *Remote Sensing of Environment*, **175**, 196–204 (2016)
2. K. Zhigalov, Usage of the game technologies engines for the purpose of modern geographic information systems creation, *ITM Web of Conferences*, **6**, 03018 (2016)
3. E. Pluzhnik; E. Nikulchev, Virtual Laboratories in Cloud Infrastructure of Educational Institutions *2014 2nd International Conference on Emission Electronics (ICEE)*, 67-69 (2014)
4. K. Zhigalov, *Cloud of Science*, **3**, 71-80 (2016)
5. In-Ae Yeoa, Jurng-Jae Yee. *Automation in Construction*, **71**, 398–413 (2016)

Quantitative Model-Free Method for Aircraft Control System Failure Detection

Eugene Zybin[1,*], *Vladislav* Kosyanchuk[1], and *Sergey* Karpenko[1]

[1]FGUP "GosNIIAS", 125319 Moscow, Russia

Abstract. The problem of the failure detection in the aircraft control system in the presence of disturbance is considered. A history based model-free nonstatistical method using the aircraft control and state data measurements only is proposed. The method needs no a priori information about the model of an aircraft, solving the prediction, identification and training problems.

1 Introduction

Faults in the aircraft control system are the most dangerous and can lead to an accident. In the event of such faults aerodynamic coefficients of the aircraft and moment characteristics of the control surfaces are changed. An important problem is to detect the abnormal dynamics of the aircraft as fast as possible.

As a rule, for the control system fault detection we use methods implying the existence of any priori information about aircraft model parameters. These methods employ three different approaches for the fault detection. The first approach is based on determining some model invariants, the second is based on solving the prediction problem, and the third is based on analytical redundancy [1–3].

In such model-based methods the parameter errors in aircraft models inevitably increase the threshold values of the fault detection criteria, thus increasing the time of the fault detection and decreasing the accuracy of determining the time the fault occurred. The derivation of error-free aircraft models proves to be practically a very hard problem [4].

The methods that do not use any priori information about the model may be qualitative or quantitative. Qualitative model-free methods are subjective analyzing the behavior of processes or employing expert systems.

Quantitative ones can be subdivided into statistical and nonstatistical methods. The statistical methods, which themselves are subject to inevitable errors, include principal component methods, partial least square methods, and methods based on classification algorithms. Determination of accurate and reliable solutions using statistical algorithms requires a large amount of data. They are characterized by high computational costs and response times.

The well-known nonstatistical quantitative model-free methods include methods based on artificial neural networks and genetic algorithms only. They require preliminary training/tuning for a particular aircraft.

The nonstatistical quantitative model-free method that does not need training is described in [5]. It uses only the control signals and data measuring of aircraft motion parameters. It's needed no a priori information about aircraft parameters and is based on an algebraic solvability condition for the problem of identifying the aircraft mathematical model. The main disadvantage of this method is its low reliability under disturbances. The paper develops this method to make it valid in case of external bounded disturbances.

2 Problem formulations

Let the model of the nonfaulted aircraft be represented in the state space as [5]:

$$x_{i+1} = Ax_i + Bu_i + m_o, \tag{1}$$

where A, B are the matrices of eigen dynamics and control efficiency; x is the state vector of length n_x; u is the output signal of the control system that, if no faults occurred, coincides with the control deflection vector of length n_u; $m_o = Bu_o$ is the vector of the constant coefficients that depend on the trim deflections of the controls; u_o is the vector of the trim deflections corresponding to the equilibrium state of the aircraft; $i = \overline{0, l-1}$ is the discrete time before the occurrence of fault; and l is the instant a fault occurs.

When fault occurs in the control system, the model of the aircraft is rewritten as

$$x^f_{j+1} = Ax^f_j + Bu^f_j + m_o, \tag{2}$$

where $j = l, l+1, ...$ is the discrete time after a fault occurs and x^f is the state vector of the faulted aircraft whose control deflection is described by the expression

$$u^f_j = Fu_j + (I - F)u^f_o, \tag{3}$$

where F is the matrix of faults (loss of efficiency) of the control system

$$F = \mathrm{diag}\left[f(1) \cdots f(k) \cdots f(n_u) \right],$$

* Corresponding author: zybin@mail.ru

u_o^f is the vector of control jamming in the case of fault

$$u_o^f = \left[u_o^f(1) \cdots u_o^f(k) \cdots u_o^f(n_u)\right]^{\mathrm{T}}.$$

Let us substitute (3) into (2) and write the model of the aircraft with the faulted control system as

$$x_{j+1}^f = Ax_j^f + B_f u_j + m_o^f, \qquad (4)$$

where $B_f = BF$ is the matrix of control efficiency for the faulted aircraft and $m_o^f = B(I-F)u_o^f + m_o$ is the constant vector characterizing the combined control deflection in the case of fault. It is required, based only on the measurements of control signals and states, to detect faults in the control system of the dynamic aircraft.

3 The deterministic problem solution

Assume that the aircraft is observed over a certain period of time. Then the aircraft models in nonfaulted (1) and faulted (4) states are written in a matrix form as $X_{i+1} = AX_i + BU_i + m_o e$, $X_{j+1}^f = AX_j^f + B_f U_j + m_o^f e$,

where $e = [1 \ldots 1]$, $X_i = [x_i \ldots x_{i+h}]$, $X_j^f = [x_j^f \ldots x_{j+h^f}^f]$, $U_i = [u_i \ldots u_{i+h}]$, $U_j = [u_j \ldots u_{j+h^f}]$; h, h^f are the numbers of the observation steps for the nonfaulted and faulted states, respectively.

The problems of identifying the model parameters of the aircraft are described by the linear right-hand matrix equations in the unknown A, B, m_o, B_f, m_o^f:

$$\begin{bmatrix} A & B & m_o \end{bmatrix} \begin{bmatrix} X_i \\ U_i \\ e \end{bmatrix} = X_{i+1}, \quad \begin{bmatrix} A & B_f & m_o^f \end{bmatrix} \begin{bmatrix} X_j^f \\ U_j \\ e \end{bmatrix} = X_{j+1}^f.$$

These equations are solvable when and only when the following conditions are satisfied [5]:

$$X_{i+1} \overline{\begin{bmatrix} X_i \\ U_i \\ e \end{bmatrix}}^R = 0, \quad X_{j+1}^f \overline{\begin{bmatrix} X_j^f \\ U_j \\ e \end{bmatrix}}^R = 0, \qquad (5)$$

where

$$\begin{bmatrix} X_i \\ U_i \\ e \end{bmatrix} \overline{\begin{bmatrix} X_i \\ U_i \\ e \end{bmatrix}}^R = 0, \quad \begin{bmatrix} X_j^f \\ U_j \\ e \end{bmatrix} \overline{\begin{bmatrix} X_j^f \\ U_j \\ e \end{bmatrix}}^R = 0.$$

Expressions (5) show that the problem of aircraft linear model identification is solvable both before and after the occurrence of fault. However, at the instant of fault occurrence behavior of the aircraft cannot be described by a single linear model. This fact is used in [5] to detect fault by a criterion that characterizes the identification problem solution accuracy:

$$\varepsilon = \left\| \begin{bmatrix} X_{i+1} & X_{j+1}^f \end{bmatrix} \overline{\begin{bmatrix} X_i & X_j^f \\ U_i & U_j \\ e & e \end{bmatrix}}^R \right\|_2, \qquad (6)$$

where matrix zero divisor of the input and output data

$$\begin{bmatrix} X_i & X_j^f \\ U_i & U_j \\ e & e \end{bmatrix} \overline{\begin{bmatrix} X_i & X_j^f \\ U_i & U_j \\ e & e \end{bmatrix}}^R = 0 \qquad (7)$$

has an orthogonal form

$$\left(\overline{\begin{bmatrix} X_i & X_j^f \\ U_i & U_j \\ e & e \end{bmatrix}}^R \right)^{\mathrm{T}} \overline{\begin{bmatrix} X_i & X_j^f \\ U_i & U_j \\ e & e \end{bmatrix}}^R = I.$$

The criterion (6) does not require a priori information about the aircraft model, solving the problems of identification and prediction while using only the measurement data and state control vectors. However, this method has a serious shortcoming. As it is based on exact equality (7), even the smallest system disturbances can lead to the essential change of structure of zero divisor. This eventually leads to low reliability procedure for detecting faults in practice.

4 The disturbed problem solution

To increase the reliability of detection of the fact and the time of occurrence of the fault in the aircraft control system in the presence of disturbance instead of an exact zero divisor (7) we will find its approximate value, a so-called numerical zero divisor. For this purpose we will write down the equation for numerical right zero divisor calculation of the some matrix C:

$$CZ \approx 0 = \Theta_{m \times s}. \qquad (8)$$

The degree of the equation (8) solution approximation can be defined by a finite small value, which characterizes the permissible level of disturbances in the system, which can be evaluated with the help of Frobenius norm, also known as the Hilbert-Schmidt or Shura norm:

$$\delta = \|\Theta\|_2 = \sqrt{\sum_{i=1}^{\min\{m,s\}} \sigma_i^2},$$

where σ_i are the singular values of the matrix Θ.

For ensuring the given norm we use the singular value decomposition of a matrix C:

$$C = C^{LT}\Sigma_C C^{RT} = \begin{bmatrix} C^{LT} & \overline{\overline{C}}^{LT} \end{bmatrix} \begin{bmatrix} \Sigma_C^{\max} & 0 \\ 0 & \Sigma_C^{\min} \end{bmatrix} \begin{bmatrix} C^{RT} \\ \overline{\overline{C}}^{RT} \end{bmatrix}, \qquad (9)$$

where C^L, C^R are the matrices of left and right singular vectors satisfying the orthogonality conditions $C^{LT}C^L = I$, $C^R C^{RT} = I$; $\Sigma_C^{\min}$ is the diagonal matrix of the minimum singular values satisfying a condition $\|\Sigma_C^{\min}\| \le \delta$; $\Sigma_C^{\max}$ is the diagonal matrix of the maximum singular values; C^L, C^R, $\overline{\overline{C}}^L$, $\overline{\overline{C}}^R$ are the matrix of left and right singular vectors corresponding to the maximum and minimum singular value.

Substituting (9) into equation (8)

$$\begin{bmatrix} \tilde{C}^{LT} & \overline{\overline{C}}^{LT} \end{bmatrix} \begin{bmatrix} \Sigma_C^{\max} & 0 \\ 0 & \hat{\Sigma}_C^{\min} \end{bmatrix} \begin{bmatrix} C^{RT} \\ \overline{\overline{C}}^{RT} \end{bmatrix} Z = \Theta$$

and premultiplying the resulting expression by the matrix of left singular vectors we do not change the norm of the right side of the equation:

$$\begin{bmatrix}\Sigma_C^{\max} & 0\\ 0 & \Sigma_C^{\min}\end{bmatrix}\begin{bmatrix}C^{RT}\\ \overline{\overline{C}}^{RT}\end{bmatrix} Z = \begin{bmatrix}C^{L}\\ \overline{\overline{C}}^{L}\end{bmatrix}\Theta = \Theta^{*}. \qquad (10)$$

Let us introduce an intermediate matrix

$$Z = \begin{bmatrix}C^{R} & \overline{\overline{C}}^{R}\end{bmatrix}\begin{bmatrix}\Upsilon\\ \Psi\end{bmatrix} \qquad (11)$$

and substitute the expression (11) into (10):

$$\begin{bmatrix}\Sigma_C^{\max} & 0\\ 0 & \Sigma_C^{\min}\end{bmatrix}\begin{bmatrix}C^{RT}\\ \overline{\overline{C}}^{RT}\end{bmatrix}\begin{bmatrix}C^{R} & \overline{\overline{C}}^{R}\end{bmatrix}\begin{bmatrix}\Upsilon\\ \Psi\end{bmatrix} = \begin{bmatrix}\Sigma_C^{\max} & 0\\ 0 & \Sigma_C^{\min}\end{bmatrix}\begin{bmatrix}\Upsilon\\ \Psi\end{bmatrix} = \Theta^{*}.$$

This expression implies that the given accuracy of the solution is provided only when $\Upsilon = 0$:

$$Z = \overline{\overline{C}}^{R}\Psi,$$

where Ψ is an arbitrary orthogonal matrix satisfying $\Psi^T\Psi = I$. Then the fault detection criterion in the presence of disturbances has a following form:

$$\varepsilon = \left\| \begin{bmatrix}X_{i+1} & X_{j+1}^{f}\end{bmatrix} \overline{\overline{\begin{bmatrix}X_i & X_j^f\\ U_i & U_j\\ e & e\end{bmatrix}}}^{R} \right\|_2, \qquad (12)$$

where the numerical zero divisor satisfies the expression

$$\left\| \begin{bmatrix}X_i & X_j^f\\ U_i & U_j\\ e & e\end{bmatrix} \overline{\overline{\begin{bmatrix}X_i & X_j^f\\ U_i & U_j\\ e & e\end{bmatrix}}}^{R} \right\|_2 \le \delta. \qquad (13)$$

Thus, to detect the fact and the time of aircraft control system fault occurrence it is necessary at each time step to define the right singular vectors of data matrices (13), corresponding to the minimum singular values with given degree of accuracy and to check the excess of a certain threshold value criterion (12).

5 The fault detection example

Let us demonstrate the validity of the proposed criterion on the example of the fault detection in the right stabilizer actuator of a highly-maneuverable aircraft [5]. Assume external disturbances $\|w\| \le 0.1$ and a fault in the form of a right stabilizer actuator jamming in the neutral position occurs in the fifth second of the flight. Fig. 1 shows the values of fault detection criterion (12) for 10 s time period.

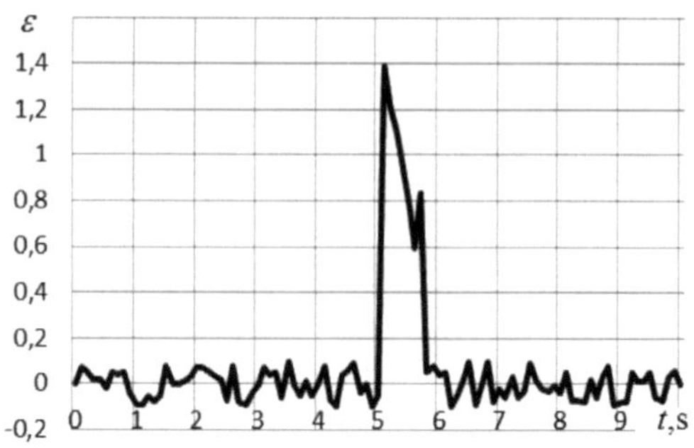

Fig. 1. The values of fault detection criterion.

It can be seen that before and after the occurrence of the fault, the values of criterion (13) is almost zero, while the fault itself is characterized by a spike the width of which corresponds to that of the identification window ($h=h^f=8$). Such a drastic change in the behavior of the plot makes it possible to accurately determine the time the fault occurred. Thus, similar to deterministic case [5] the fault is detected in the shortest possible time, corresponding to the integration step (0.1 s).

6 Conclusions

In this article the new quantitative model-free method for aircraft control system failure detection in the presence of external disturbances is proposed. It does not depend on the model parameters and is based only on the information about the observed signals involving no other auxiliary variables. This ensures its validity in the absence of any priori information without any training, identification or predicting procedures.

References

1. X. Qi, D. Theilliol, J. Qi, Y. Zhang, J. Han, D. Song, L. Wang, Y. Xia. Fault diagnosis and fault tolerant control methods for manned and unmanned helicopters: a literature review. *Proc. of the Conference on Control and Fault-Tolerant Systems SysTol*, pp. 132–139 (2013)
2. R.J. Patton, Fault-tolerant control: the 1997 situation. *Proc. of the IFAC Symposium on Fault Detection Supervision and Safety for Technical Processes*, pp. 1033–1055 (1997)
3. Y. Zhang, J. Jiang. Ann. Rev. Control. **32**(2), 229–252 (2008)
4. O.N. Korsun, *Mekhatronika, Avtomatizatsiya, Upravlenie*. **S6**, 2–7 (2008)
5. E.Yu. Zybin, V.V. Kosyanchuk. *Journal of Computer and Systems Sciences International*, **55**(4), 546–557 (2016)

AUTHOR INDEX